Silicon, From Sand to Chips 2

Silicon, From Sand to Chips 2

Microelectronic Chips, Solar Cells, MEMS

Alain Vignes

WILEY

First published 2024 in Great Britain and the United States by ISTE Ltd and John Wiley & Sons, Inc.

ISTE Ltd
27-37 St George's Road
London SW19 4EU
UK

www.iste.co.uk

John Wiley & Sons, Inc.
111 River Street
Hoboken, NJ 07030
USA

www.wiley.com

Any opinions, findings, and conclusions or recommendations expressed in this material are those of the author(s), contributor(s) or editor(s) and do not necessarily reflect the views of ISTE Group.

Library of Congress Control Number: 2023950584

British Library Cataloguing-in-Publication Data
A CIP record for this book is available from the British Library
ISBN 978-1-78630-922-8

Contents

Preface

At the beginning of the 20st century, silicon "metal" was used as an alloying element for steels with electrical properties. The year 1906 saw the first application of crystalline silicon as a component of electromagnetic wave detection circuits in radio receivers, competing with galena.

Research carried out during the Second World War on silicon and germanium, the materials used in the components (point-contact diodes) of radar receiver circuits for aircraft detection and tracking, revealed that these materials are semiconductors whose basic characteristic is the control of electrical conductivity through doping. This characteristic prompted the search, after the Second World War, for solid components to replace "triodes" (vacuum tubes). This quickly led to the invention of the transistor.

The invention of the transistor is the founding act of the digital revolution (of the information society in which we live).

Germanium then silicon are the first two materials that enabled the invention of the transistor and the initial development of computers, while silicon dethroned germanium to produce the "MOSFET" (metal–oxide–semiconductor field-effect transistor), the basic component of integrated circuits: microprocessors and memories, the building blocks of computers.

But these components require materials (germanium and silicon) of extraordinary purity and perfect crystallinity. The purification of basic materials to purities of up to 11N, the production of single crystals of germanium, then silicon, the manufacture of components (based on transistors) and their miniaturization have posed problems of a complexity rarely encountered in the development of manufactured products.

These are the same properties and characteristics that have made silicon the material of choice for converting solar energy into electricity and for photographic sensors.

Silicon's exceptional mechanical properties, combined with its electrical properties, make it the material of micro-electro-mechanical systems (MEMS), the key components of "intelligent objects".

In 2018, there were no materials on the horizon that were likely to dethrone silicon as the material of choice for microelectronics and optoelectronics alike. According to Gérard Berry: "Silicon is not dead, far from it".

This book is aimed at readers who want to know and understand how it was possible to go from the ENIAC computer, built during the Second World War, to calculate shell trajectories, 30 m long and 2 m high, with 17,468 triodes (vacuum tubes) and capable of executing 5,000 additions and subtractions in 1 s, to centimetric microprocessors with 20 billion transistors, processing power (number of instructions processed per second) of several gigahertz, making up the basic components of the individual computer, which is the size of a thin book.

To this end, this book, by tracing the history of discoveries, inventions, innovations and technological developments in materials, components, integrated circuits and memories, presenting the physical bases of their operation, and focusing on the materials and technologies used to make these components, attempts to answer the following questions:

– What specific properties (characteristics) – electrical, physicochemical, mechanical – are behind the successive dominance of silicon, then germanium, then silicon again in the development of microelectronics, the dominance of silicon in the conversion of solar energy into electricity, the dominance of silicon as the basic material for electromechanical microsystems?

– What properties (purity, crystallinity, doping) had to be imparted to the material, and how were they obtained to achieve the performance achieved by these components today?

– What processes had to be developed to produce these components, and then to meet the demands of miniaturization, enabling the high-speed data processing performance we are seeing today, efficient conversion of solar energy into electricity, etc.?

– Who were the architects of this epic? According to Gérard Berry[1], "its extraordinary success (that of silicon) is clearly to the credit of semiconductor materials physicists, who made technological advances that required enormous imagination and skill to overcome all the obstacles".

Until 1942, silicon extracted from silica (SiO_2) and germanium extracted from sulfide (GeS_2) were considered as metals. The semiconductors known at the time were chemical compounds: oxides (Cu_2O) and sulfides (galena PbS), composed of a metal and a metalloid (oxygen or sulfur), whose basic characteristic was the increasing variation of their conductivity with temperature, whereas the conductivity of metals decreases with increasing temperature. It was not until the summer of 1942 that it was recognized that purified silicon and germanium were not metals, but semiconductors.

This book is divided into two volumes. Volume 1 is devoted to basic components (diodes and transistors).

Chapter 1 presents (1) the work that led to the extraction of silicon from silica and its purification and the discovery, extraction and purification of germanium; (2) the basic physical characteristics of semiconductors made from these two materials, knowledge of which is essential for understanding how components work.

Chapters 2–6 of Volume 1 present the basic components (diodes, transistors) in the chronological order of their discovery/invention, and the technological developments required for their realization.

Each chapter includes a presentation of the component, how it works and its basic functions, followed by the history of the research and development that led to its invention and production. The physical basis of its operation is presented in the appendicies of each chapter. The technologies used to satisfy the requirements of purity and crystalline perfection of the base material are presented chronologically, as are the technologies used to produce the components and the evolutions required by their miniaturization. The industrial development of the first components is presented according to their importance for subsequent developments.

Volume 2 is devoted to "chips, optoelectronic components and MEMS".

Chapters 1 and 2 present microcomputer integrated circuits and memories.

Chapter 3 presents the silicon thin film transistor TFT, which led to the development of flat-panel liquid crystal displays.

---

1 Berry, G. (2017). *L'Hyperpuissance de l'informatique*. Odile Jacob, Paris, p. 88 and 401.

Chapters 4 and 5 present silicon optoelectronic components. These include solar cells for converting solar energy into electricity and photoelectric image sensors for digital cameras, which have revolutionized astronomy and medical imaging.

Chapter 6 presents microelectromechanical systems (MEMS), the exceptional mechanical properties of silicon that have enabled their development, and the specific technologies developed for building structures with moving parts.

Many English and American books present the "history of semiconductors". Compared with the reference works cited in the reference lists, this book presents not only the historical aspects, but also the recent technological developments that have enabled the current performance of microprocessors, memories, solar cells and electromechanical microsystems. The book is based on numerous works by historians and original publications.

The author would particularly like to thank Professors Jean Philibert and André Pineau.

January 2024

References

Burgess, P.D. (n.d.). Transistor history [Online]. Available at: https://sites.google.com/site/transistorhistory.

Computer History Museum (n.d.). The silicon engine timeline [Online]. Available at: www.computerhistory.org.

Hu, C. (2009). *Modern Semiconductor Devices for Integrated Circuits*. Pearson, London.

Krakowiak, S. (2017). Éléments d'histoire de l'informatique. Working document, Université Grenoble Alpes & Aconit, CC-BY-NC-SA 3.0 FR.

Lazard, E. and Mounier-Kuhn, P. (2022). *Histoire illustrée de l'informatique*. EDP Sciences, Les Ulis.

Lilen, H. (2019). *La belle histoire des révolutions numériques*. De Boeck Supérieur, Louvain-la-Neuve.

Lojek, B. (2007). *History of Semiconductor Engineering*. Springer, New York.

Mathieu. H. (2009). *Physique des semiconducteurs et des composants électroniques*, 6th edition. Dunod, Paris.

Nouet, P. (2015). Introduction to microelectronics technology. Working document, Polytech Montpellier, ERII4 M2 EEA Systèmes Microelectronics.

Orton, J.W. (2004). *The Story of Semiconductors*. Oxford University Press, Oxford.

Orton, J.W. (2009). *Semiconductors and the Information Revolution: Magic Crystals that made IT Happen.* Elsevier, Amsterdam.

Riordan, M. and Hoddeson, L. (1997). *Crystal Fire: The Invention of the Transistor and the Birth of the Information Age.* W.W. Norton & Company, New York.

Seitz, F. and Einspruch, N.G. (1998). *Electronic Genie: The Tangled History of Silicon.* University of Illinois Press, Illinois.

Sze, S.M. (2002). *Semiconductor Devices: Physics and Technology.* Wiley, New York.

Verroust, G. (1997). Histoire, épistémologie de l'informatique et révolution technologiques. Course summary, Université Paris VIII, Paris.

Ward, J. (n.d.). Transistor museum [Online]. Available at: transistormuseum.com.

Introduction

The Digital Revolution

The "digital revolution" is also known as the "computing or IT revolution". These expressions reflect "a radical transformation of the world we are witnessing today".

The first term refers to the binary digitization of texts and numbers, as well as images, sounds and videos, using sequences of symbols. This makes it possible to store images, sounds, etc., and transmit them, replicate them, analyze them and transform them using digital computers (Abiteboul and Dowek 2017, p. 29).

The second expression, "the computing revolution", refers to the science and technique of processing digitized information using algorithms. According to Berry (2017, p. 25), "Computing is the conceptual and technical engine of the digital world. The computer is the physical engine".

The "birth certificate of the digital revolution" is Claude Shannon's 1937 master's thesis, *A symbolic analysis of relay and switching circuits* (1938). This thesis relied on the theory of the Englishman George Boole (*An Investigation of the Laws of Thought*, 1847), which established the link between calculus and logic and where the basic logical functions "AND", "OR" and "NOT" were treated as arithmetic operations, taking the value 0 or 1, depending on whether the proposition was true or false.

The master's thesis of Claude Shannon[1] was the result of an internship at Bell Labs[2], where he observed the power of telephone exchange circuits that used

---

1 Claude Shannon is also the father of the information theory formulated in 1950, described by the journal *Scientific American* as the "Magna Carta of the information age". (Collins 2002; Berry 2017, p.52).

2 Bell Labs: Bell Telephone Laboratories, a subsidiary of ATT (American Telephone and Telegram Company) with a monopoly on telephone and telegraph transmissions. Its subsidiary, Western Electric, produces components for telephone exchanges.

electromechanical relays (switches)[3] to route calls and imagined that electrical circuits could perform these logical operations using an on-off switch configuration.

The first demonstration of the feasibility of executing logic functions using a device made up of two electromechanical relays was carried out in 1937 by George Stibitz of Bell Labs; this led to the construction in 1939 of the first CNC (complex number calculator) (400 electromechanical relays), capable of opening and closing 20 times a second, executing complex number multiplication and division operations. This was followed by five other models. "Stibitz's calculator demonstrated the potential of a relay circuit to do mathematics in binary, process information, and manipulate logical procedures" (Isaacson 2015, p. 93).

The "digital" revolution is the third major revolution in human history. The first was the agricultural revolution 8,000 years ago. The second was the "industrial revolution" of the 19th century.

The technology at the heart of this third revolution, also known as the "second industrial revolution", is microelectronics[4]. In 1979, the US National Academy of Sciences published a report[5] entitled "Microstructure, Science, Engineering and Technology", which stated: "The modern era of electronics has ushered in a 'second' industrial revolution, the consequences of which may be even more profound than those of the first". According to Ian Ross, President of Bell Labs from 1979 to 1991: "The semiconductor odyssey produced a revolution in our society at least as profound as the total industrial revolution. Today electronics pervades our lives and affects everything" (Ross 1997).

I.1. Microelectronics components

In 1903, Arthur Fleming invented the diode (vacuum tube), a current rectifier, and in 1906, Lee de Forest invented the triode (vacuum tube) by adding a grid between the diode's cathode and anode. As well as rectifying the current, this allowed weak currents induced by electromagnetic waves to be amplified, hence the development of

---

3 The electromechanical relay (a switch that opens and closes by electrical means, such as an electromagnet), with its two states open and closed, was the ideal component for representing the two states of binary numbering (0 and 1) and logic (true or false). With a binary machine, arithmetic operations and logical operations can be processed in the same way.

4 Microelectronics refers to all the technologies used to manufacture components that use electrical currents to transmit, process or store information.

5 Quoted in the brochure "La microélectronique : bilan et perspectives d'une technologie de base", Siemens Aktiengesellschaft, Berlin and Munich, 1984. Translation of the book *Chancen mit Chips*, 1984.

radio receivers: a small variation in the signal on the grid resulted in an amplification of the cathode-anode current. In addition, a sudden variation in the signal applied to the grid switched the triode on or off, enabling it to function as a switch. The triode is also capable of self-oscillation, hence its use in radio transmitters.

The invention of the bipolar transistor in 1948 by William Shockley (Nobel Prize winner), a solid-state device capable of performing the same functions (amplification of weak currents and switching), but much faster, ushered in the era of the digital revolution.

Like transistors, triodes work by controlling a current of electrons, which can either be amplified or interrupted and reignited. These components function like a switch that can be set to 0 or 1 on command, thus performing logic functions. But with triodes, switching times are much longer and the permissible frequencies much lower than in solid-state components, because these variables are linked to the time taken for the electrons to cross the distance between the cathode and the anode (around 1 mm); whereas, in a transistor, the distance traveled by the electrons between the emitter and the collector is less than 1 μm, down to around 20 nm.

Before the invention of the transistor, prototype "computers" had been built with triodes, the ENIAC during the Second World War, then with solid diodes (made of germanium) combined with triodes. Diodes can only be used to create logic circuits (OR and AND gates). They cannot restore the signal at the output of a gate, hence the presence of triodes to restore the signal, enabling cascades of gates to be created, and hence logic circuits.

The discovery of silicon N and silicon P[6], at the beginning of the Second World War, in other words of the effect of doping on the conductivity of silicon and therefore its control, and the discoveries of the rectifier effect and the photoelectric effect presented by the solid-state PN diode[7] by Rüssel Ohl, led to the invention of the bipolar transistor (with PN junctions) in 1949 by William Shockley. The development of circuits made up of solid-state diodes and transistors producing NAND and NOR logic gates and all the universal logic functions by combining one or the other, with the added feature of restoring the signal at the output of each gate, thus enabling cascades of logic gates, led to the development of integrated circuits, invented in 1958–1959 by Jack Kilby (Nobel Prize winner) and Robert Noyce.

The development of the silicon-based field-effect MOSFET transistor, designed by William Shockley in 1945 and by Dawon Kahng and Martin Atalla in 1960, because

---

6 Silicon N (silicon doped with phosphorus) with n-type conductivity; P silicon (doped with boron) with p-type conductivity.

7 PN diode: a component made up of two regions, N and P, joined together along a flat surface.

of its miniaturization capacity, enabled the development of integrated circuits: memories and microprocessors. Microprocessors were the ultimate innovation in the digital revolution, enabling the development of the personal computer. According to Reid (1984), "A new era in electronics had begun".

The miniaturization of components down to the nanometer scale is delivering high performance in terms of information processing speed and substantial savings in power consumption. The number of transistors has risen from 2,400 for the Intel 4004, the first integrated microprocessor, to around 20 billion for today's largest graphics processors (2017). The processing power of a microprocessor (the number of instructions a microprocessor is capable of processing per second) rose from a few megahertz in the early 1980s to several gigahertz in the early 2000s. This clock frequency (as it is known) is directly linked to the switching speed of the microprocessor transistors. We can only marvel that an astronomical set of phenomenally fast electronic components as simple as switches could be the basis of humanity's third revolution.

I.2. Microelectronics materials

These "components" require materials of extraordinary purity and perfect crystallinity to obtain very specific electronic characteristics, as well as completely new technologies for manufacturing transistors and integrated circuits (a list of which is given in the Appendix).

It was the availability and technological mastery of two materials, germanium and silicon, which were virtually unknown at the beginning of the 20th century, with the appropriate electronic characteristics, that enabled the invention of the transistor and conversion of solar energy into electricity.

The purification of base materials to purities of up to 11N (99.999999999), the production of perfectly crystalline single crystals of germanium and then silicon, enabling the conductivity of these materials to be controlled by doping, and the manufacture of components and their miniaturization have posed problems of a complexity rarely encountered in the development of manufactured products (Queisser 1998).

It was with germanium (on purified, coarse-grained (quasi monocrystalline) wafers that were available) that power amplification was first observed in December 1947 on a device made by John Bardeen and Walter Brattain (Nobel Prize winners), which was named the "point contact transistor" (Bardeen and Brattain 1948). This invention led to the development of a process for obtaining single crystals of germanium by Teal (1976). The successful purification and manufacture of germanium single crystals and the development of the bipolar transistor established germanium as the basic material for transistors. In 1952, Ralph Hunter, in a speech as President of the Electrochemical Society of the United States, predicted: "A revolution in the electronics industry as a result of the development of germanium". Germanium transistors were manufactured until 1961. The CDC 1604 and IBM 1401 computers marketed in 1960 were made using germanium transistors.

In 1952, following the successful manufacture of silicon single crystals and of a PN junction in a single crystal, again by Gordon Teal, whose properties were superior to those of the germanium PN junction, "silicon immediately became a rival to germanium" (Leamy and Wernick 1997). Given the difficulties in obtaining "electronic" silicon, silicon very gradually became the preeminent material for transistors, under pressure from the military, who were virtually the only customers at the time – particularly for the temperature resistance of silicon diodes and transistors up to around 150°C.

When the first silicon MOSFET transistor was produced in 1959, silicon's supremacy became total, thanks to the qualities of its oxide and its high thermal dissipation. Since the 1970s, silicon MOSFETs have been the basic components of integrated circuits and computer memories.

In 1951, Heinrich Welker (Nobel Prize) began studies on compounds with the same structure as silicon and germanium, such as gallium arsenide GaAs, revealing their semiconductor characteristics. It was not until 1978 that it was shown that a gallium arsenide component was twice as fast as the same silicon component under the same conditions (Welker 1976). Nevertheless, this factor of 2 did not convince manufacturers to abandon silicon, thanks to its two advantages: its high heat dissipation and its mastered technology (Bols and Rosencher 1988).

I.3. The driving forces behind the development of microelectronics and computer components

Research studies carried out in England and the United States from the start of the Second World War on the reception of radar electromagnetic waves by the "point contact diode" were the first driving force behind the development of silicon and germanium, and marked the first victory of this solid component over vacuum tubes.

The second driving force behind the development of microelectronics was the desire of Bell Labs[8], from the end of the war, to find a solid substitute for the

---

8 "The research group established at Bell Labs in the summer of 1945 had a long-term goal of creating a solid state device that might eventually replace the tube and the relay" (Ross 1997).

triode lamps used as amplifiers along telephone transmission lines and for the electromechanical relays in their ATT telephone exchanges[9].

It was the discoveries of silicon N and silicon P, of the property of rectifying an electric current through a unidirectionally solidified silicon ingot, constituting a PN diode, and of the photovoltaic effect presented by this ingot in 1940, that were at the origin of Bell Labs' adventure in microelectronics. When these remarkable properties of a silicon ingot were brought to the attention of the Bell Labs director, Mervin Kelly considered this discovery of great value to the electronics industry, and decided that absolute secrecy should be preserved until in-depth studies revealed its full power: "It was too important a breakthrough to bruit about".[10] The studies were resumed in 1945.

In the summer of 1945, as reported by Ian Ross, Kelly set up a research group with the following objectives: the fundamental study of semiconductors, concentrating on germanium and silicon, materials which were beginning to be well known, and, in the long term, the creation of a solid-state component constituting an amplifier "to replace triodes (vacuum tubes) and constituting a switch to replace the electromechanical relays of telephone exchanges".

This research, in 1947 and 1949, led to the invention of the point contact transistor and the bipolar transistor with PN junctions.

In 1950, according to Ian Ross, Bell Labs researchers realized that, given the characteristics of transistors, their size and low energy consumption, it was not the replacement of vacuum tubes that should be sought, *but their use as components of logic circuits*[11].

---

9 In the 1950s, a great deal of research was undertaken to apply vacuum tubes to telephone exchange switching. The results were unsuccessful. Until 1974, semi-electronic switches ("space switching") were installed: hybrid switches whose control system is entirely electronic, but which operate on a connection network that is still mechanized, circulating purely analogue conversation currents. The first experimental telephone exchange to use an entirely electronic switching system based on microprocessors, known as "time or digital switching", was set up by the CNET in Perros-Guirec, France, in 1970. These switches constitute the real revolution in modern telecommunications (Caron 1997, p. 295).

10 "The goal for the group, following Mervin Kelly's instructions, was to determine whether it was possible to develop a practical semiconductor triode" (Seitz and Einspruch 1998, p. 164).

11 Ross quoted Bob Wallace: "Gentlemen, you've got it all wrong. The advantage of the transistor is that it is inherently a small size and a low power device. This means that you can pack a large number of them in a small space without excessive heat generation and achieve low propagation delays. And that is what we need for logic applications. The significance of the transistor is not that it can replace the tube but that it can do things that the vacuum could never do". And according to Ross (1997), "And this was a revelation to us all".

As soon as the reproducible manufacture of transistors became possible, in the mid-1950s, "replacing vacuum tubes in as many applications as possible became the objective".

Transistor specimens were entrusted to various Bell Labs engineers with the task of developing applications. John H. Felker was one of them. Felker (1951) showed that the transistor could be used as a component of logic circuits. This potential use of the transistor was presented by Felker to the companies that had acquired the "Western Electric" license. According to McMahon[12], "none of us imagined the revolution that would take place over the next forty years", "even at IBM" according to Rick Dill[13].

Following this presentation, in 1951, the Air Force asked Bell Labs to develop a computer, the TRADIC (transistorized airborne digital computer), which was entrusted to Felker. This resulted in the successive production of four TRADIC computers, of which the Leprechaum version, operational in 1956, was the first fully transistorized computer based on logic circuits made up of bipolar germanium transistors (Irvine 2001).

Most of the discoveries, inventions and technological developments relating to transistors, solar cells and digital photography between 1947 and 1970 were made by Bell Labs researchers (see Table I.1 in the Appendix in this chapter).

Nevertheless, as we shall see, the inventions and technological developments of Bell Labs were not always followed by industrial development and production by Western Electric. There was a good reason for this: ATT, which had a monopoly over telephone and telegraph transmissions in the United States by court order under the anti-trust laws, was only authorized to produce electronic components for its own needs and had to inform the entire electronics industry of any discoveries that might be of interest to it. Therefore, after the first bipolar transistor was produced in 1950, Western Electric began to grant manufacturing licenses to companies producing diodes, triodes (vacuum tubes), etc., "licensing the rights to manufacture transistors for a $25,000 fee", and for these licensees, Bell Labs organized a Transistor Technology Symposium in April 1951.

---

12 "First, J.H. Felker fascinated us with the applicability of transistors to high-speed digital computers. He stated that the prime objective was practically infinite reliability, closely followed by low power consumption (hence minimal heat removal), small size and minimal weight. However, none of us imagined the semiconductor revolution that was really to take place over the next forty years" (McMahon 1990) (Hughes Aircraft Company).
13 "Everyone on the early transistor business saw analog and communications circuits as the most important thing. In 1954, computers were not important to the electronic world", Rick Dill (IBM) (Dill 1954).

The third driving force was the interest shown by a number of industrial companies who foresaw the importance of these inventions. This was as early as 1948, with the publication of the discovery of the point contact transistor, companies that had been heavily involved in the development and production of germanium diodes during the Second World War: General Electric, Sylvania, RCA, CBS and IBM. Subsequently, other vacuum tube manufacturers who had acquired the Bell license, such as Raytheon, Philco, Telefunken and Siemens, and companies set up by researchers or engineers, who moved from one company to another, produced components and then integrated circuits. In 1958, there were 70 diode and transistor manufacturers in the world, the vast majority of these in the United States (Morton and Pietenpol 1958).

The first commercial computers to use bipolar transistors as logic circuit components appeared in 1956 with the Philco S-2000 and 2600 computers.

Three companies – Texas Instruments (TI), founded in 1952, Fairchild Semiconductor, founded in 1957, and Intel, founded in 1968 – took over from Bell Labs, both in the development and industrial production of the ultimate components.

The inventions and achievements of the integrated circuit in 1958–1959 were due to Jack Kilby of Texas Instruments and Robert Noyce of Fairchild Semiconductor and their collaborators. This invention paved the way for the creation of the "microprocessor" by Intel, a company founded by Fairchild Semiconductor defectors Noyce, Grove and Moore (author of "Moore's Law" in 1969). This was a universal integrated circuit that integrated all the functions of a computer's central processing unit, capable of following programming instructions. In November 1971, Intel presented the Intel 4004 microprocessor.

At the same time, another major driving force behind the development of microelectronics, as with many other major innovations, was the needs of the military or prestige of the state.

The civil space program and the military program to build balistic missiles boosted demand for transistors. The state organizations responsible for these programs financed the companies mentioned above.

The Polaris sea-to-ground ballistic missile program in 1956, then the Minuteman ground-to-ground missile program at the end of the 1950s, for their on-board guidance system, the Vanguard and Explorer earth satellites, launched in 1958. For their transmissions, the Apollo program, at the beginning of 1960, endowed with 25 billion dollars, gave a real boost to research into integrated circuits and computers[14].

I.4. The material of solar energy

Diodes made of silicon, germanium and other semiconductors convert photons into electrons.

The photovoltaic effect, presented by a silicon ingot forming a PN diode, was discovered by Russell Ohl of Bell Labs in 1940 (Ohl 1946).

The solar cell development program began in 1952. On April 26, 1954, Bell Labs announced the manufacture of silicon solar cells using the diffusion doping process. It was the development of this doping process for solar cells that ensured the development of transistors (Chapin et al. 1954).

The space program was the driving force behind the development of solar cells; the first use of solar cells was on the Vanguard 1 satellite, launched on March 17, 1958, to power a radio transmitter. The system operated for 8 years. The space program stimulated (financed) a great deal of research and a veritable cell production industry.

The energy crisis of 1974–1975 sparked renewed interest in solar cells. Silicon is the material of a major energy source (solar energy): 99.4% of solar panels are based on silicon, and 0.4% on CdTe and GaAs.

Solar Impulse 2, the fragile aircraft with its huge wings covered with solar panels (11,628 ultra-fine monocrystalline silicon photovoltaic cells (each 135 μm thick)), is the symbol of the progress made in just a few years in the field of materials and renewable energies.

I.5. The material of digital image sensors

Silicon photoelectric image sensors, whose invention by W.S. Boyle and G.E. Smith, also of Bell Labs, in 1970 won them a Nobel Prize, have made digital

---

14 "The decision of President Kennedy in 1961 to mount an intensive space programme, with the in-tention to put a man on the moon in 1970 kick-started a technological revolution, certainly no other country ever received a comparable boost. Given the modest lifting capability of current US rockets weight was a vital factor and all electronics must therefore be transistorized. Ruggedness and relia-bility too were better served by solid-state devices than by the older fragile vacuum tubes. It became clear that the required rocket guidance would demand highly sophisticated computer technology and that such advanced circuitry could only be realised in integrated form" (Orton 2004, p. 99).

photography possible and revolutionized astronomy. They are crucial components of fax machines, cameras, scanners and medical imaging (Boyle and Smith 1970).

I.6. The material of micro-electro-mechanical components (MEMS)

A MEMS "sensor" or "actuator" is an essential part of what we call intelligent objects, since it is thanks to them that we can obtain information linked to our environment and vice versa. A series of technological breakthroughs and industrial bets have helped to explode a market that continues to evolve (Vigna 2013).

The development of these microsystems has been made possible by the availability of a material, silicon, with its exceptional electrical and mechanical properties, and by the development of specific miniaturization technologies for this material.

I.7. The role of "metallurgists"

While the invention of the transistor can undoubtedly be attributed to three physicists: John Bardeen, Walter H. Brattain, William B. Shockley of Bell Labs, according to Jack Scaff, director of the transistor laboratory materials from Bell Labs: "The role of metallurgists in these developments was essential" (Scaff 1970).

The purification of basic materials, down to purities of 11N, and the miniaturization of components (transistors) have posed material problems of a complexity rarely encountered in the development of manufactured products.

The appended table lists the main inventions, discoveries and achievements by "metallurgists/chemists"; for instance, the discovery of silicon N and silicon P, the discoveries of the rectifier effect and of the phoelectric effect, presented by a unidirectionally solidified silicon rod by Rüssel Ohl from ATT's Bell Labs in 1940.

It is to Gordon Teal of Bell Labs that we owe, from the discovery of the point contact transistor in December 1947, the recommendation to purify the material and to use single crystals as the base material of transistors, the development of the CZ pulling method to obtain single crystals of germanium initially, then of silicon, the development of processes to obtain the greatest purity of the base material, the development of the doping process for germanium and then silicon, which led to the production of the first bipolar transistor in germanium.

The work of Jack Scaff and Henry Theuerer, and then Calvin Fuller on the diffusion doping process is noteworthy; this became the basic manufacturing process for transistors (later replaced by ion implantation, based on the same principle) (Chapter 5).

The discovery of the masking process by oxidation of silicon, by Derrick and Frosch, enabled the invention, by Jean Hoerni of the Fairchild company, of the bipolar transistor with a planar configuration in silicon. According to Bo Lojek, Hoerni carried out his experiments by working alone, practically at night, without any research budget, taking care not to inform Gordon Moore (Lojek 2007, p. 123).

Atalla's discovered silicon passivation by oxidation, which led to the development of the first silicon MOSFET transistor by Khang and Atalla (Chapter 6).

I.8. The technological keys to the digital revolution

The technological keys to the digital revolution are as follows:

– The discovery of silicon N and silicon P in 1940; in other words, the discovery that the conductivity of silicon, depending on doping with elements such as boron or phosphorus, could be of the n type (by electrons carrying a negative charge) or the p type (by holes, carrying a positive charge), a property also observed for germanium. Further research in 1942 established that these elements were "semi-conductors".

– The discovery of the silicon-based PN diode in 1940.

– The invention of the bipolar transistor in 1949 by W. Shockley solid electronic component, designed to replace the triode as an amplifying element, but also proved to be a switching element: a "real electronic valve".

– The invention of the integrated circuit in 1959: a major innovation consisting of the "monolithic integration", in a single crystal of silicon, of logic circuits made up of several transistors, up to billions, which ushered in an industrial revolution (the second or third).

– Purification to unprecedented levels for a material (up to 11N) and the production of perfectly crystalline single crystals for germanium and then for silicon, enabling the conductivity of these semiconductor materials to be controlled by doping.

– The qualities of silicon oxide: stability, insulator (dielectric), diffusion barrier, selective etching by HF (Hydrofluoric acid) (without attacking the underlying silicon), which enabled the creation of the MOSFET field effect transistor and the development of integrated circuits.

– The thermal dissipation of silicon (the thermal conductivity of silicon is twice that of germanium and three times that of gallium arsenide), enabling advanced integration.

I.9. Appendix

Innovations	Companies	Dates	Authors
Controlling Si conductivity by doping	BTL	1942	Russell Ohl and Jack Scaff
Rectifier effect and photovoltaic effect of the silicon PN diode	BTL	1940	Russell Ohl
Point contact transistor	BTL	1947	John Bardeen and Walter Brattain (Nobel)
Bipolar transistor (with PN junctions)	BTL	1948	William Shockley (Nobel)
Germanium (Ge) single crystal	BTL	1950	Gordon Teal
***Grown junction* Ge bipolar transistor**	BTL	1950	Morgon Sparks and Gordon Teal
Ge purification by zone melting	BTL	1951	W.C. Pfann
Ge bipolar transistor made by alloying (*alloy junction transistor*)	General Electric	1951	R.N. Hall and W.C. Dunlap
Silicon bipolar transistor	BTL-TI	1954	Morris Tannenbaum and Gordon Teal
Production of a Si solar cell by boron diffusion	BTL	1954	Calvin S. Fuller
Oxide masking process of Si (*oxide masking*)	BTL	1955	Lincoln Derrick and Carl Frosch
Si planar bipolar transistor	Fairchild	1959	Jean Hoerni
***Oxide surface* passivation of silicon**	BTL	1959	Martin Atalla, Morris Tannenbaum and E. Scheibner
Integrated circuits	TI and Fairchild	1959	Jack Kilby (Nobel) and Robert Noyce
Si MOSFET transistor	BTL	1960	Dawon Khang and Mohamed M. Atalla
Complementary MOSFET transistor (CMOS)	Fairchild	1963	Frank Wanlass, Chih Tang Sah, Moore
***Floating gate MOSFET* (ROM memory)**	BTL	1967	Dawon Kahng and Simon Min Sze
DRAM memory (*one transistor DRAM cell*)	IBM	1968	Robert Dennard
CCD digital image sensors	BTL	1970	Willard S. Boyle and George E. Smith (Nobel)

(BTL: Bell Telephone Laboratories, "Bell Labs"; TI: Texas Instruments)

Table I.1. *Main discoveries, inventions and innovations (adapted from P. Seidenberg, "From germanium to silicon", ethw.org/archives)*

I.10. References

Abiteboul, S. and Dowek, G. (2017). *Le temps des algorithmes*. Le Pommier, Paris.

Bardeen, J. and Brattain, W.H. (1948). The transistor, a semi-conductor triode. *Physical Review*, 74(7), 230–231.

Berry, G. (2017). *L'Hyperpuissance de l'informatique*. Odile Jacob, Paris.

Bols, D. and Rosencher, E. (1988). Les frontières physiques de la microélectronique. *La recherche*, 203, 1176–1186.

Boyle, W.S. and Smith, G.E. (1970). Chargecoupled semiconductor device. *Bell System Technical Journal, B.S.T.J. Briefs*, 49(4), 587–593.

Caron, F. (1997). *Les deux révolutions industrielles du XXème siècle*. Albin Michel, Paris.

Chapin, D.M., Fuller, C.S., Pearson, G.L. (1954). A new silicon p-n junction photocell for converting solar radiation into electric power. *Journal of Applied Physics*, 25(5), 676–677.

Cohen, D. (2015). *Le monde est clos et le désir infini*. Albin Michel, Paris.

Collins, G.P. (2002). Claude E. Shannon: Founder of the information theory. *Scientific American*.

Dill, R. (1954). Germanium alloy transistors [Online]. Available at: http://ibm-1401.info/germaniumalloy.html.

Felker, J.H. (1951). The transistor as a digital computer component. In *Proceedings AIEE-IRE Computer Conference*, Philadelphia, American Institute of Electronic Engineers (February 1952), 105–109.

Irvine, M.M. (2001). Early digital computers at Bell Telephone Laboratories. *IEEE Annals of the History of Computing*, 23(3), 21–42.

Isaacson, W. (2015). *Les innovateurs*. JC Lattès, Paris.

Leamy, H.J. and Wernick, J.H. (1997). Semiconductor silicon: The extraordinary made ordinary. *MRS Bulletin*, 47–55.

Lojek, B. (2007). *History of Semiconductor Engineering*. Springer, Berlin.

McMahon, M.E. (1990). The great transistor symposium of 1951, reprinted from SMEC. *Vintage Electrics*, 2. Prologue.

Morton, J.A. and Pietenpol, W.J. (1958). The technological impact of transistors. *Proceedings of the IRE*, 6, 955–959.

Ohl, R.S. (1946). Light sensitive electric device. Patent, US2402662.

Orton, J.W. (2004). *The Story of Semiconductors*. Oxford University Press, Oxford.

Queisser, H.J. (1998). Materials research in early Silicon Valley and earlier. Semiconductor silicon. *The Electrochemical Society Proceedings Series*, PV 98-1, 4.

Reid, T.R. (1984). *The Chip*. Simon & Schuster, New York.

Ross, I.M. (1997). The foundation of the silicon age. *Bell Labs Technical Journal*, 2(4), 3.

Scaff, J.H. (1970). The role of metallurgy in the technology of electronics materials. *Metallurgical Transactions*, 1(3), 561–573.

Seitz, F. and Einspruch, N.G. (1998). *The Tangled History of Silicon*. University of Illinois Press, Illinois.

Shannon, C. (1938). A symbolic analysis of relay and switching circuits. *Transactions of the American Institute of Electrical Engineers*, 57(12), 713–723.

Teal, G.K. (1976). Single crystals of germanium and silicon – Basic to the transistor and integrated circuit. *IEEE Transactions on Electron Devices*, 23(7), 621–639.

Vigna, B. (2013). La revolution des MEMS, Interview (ST microelectronics). *Paris Tech Review*, 9 December [Online]. Available at: www.paristechreview.com.

Welker, H.J. (1976). Discovery and development of III-V compounds. *IEEE Transactions on Electron Devices*, 23(7), 664–674.

Integrated Circuits

The invention of the integrated circuit, in 1959, consisting of the implementation in a semiconductor monocrystal of logic circuits, analog circuits and memories made up of several transistors, up to billions of transistors, was the founding act of the technological revolution (microelectronics) that enabled the digital revolution.

The victory of integrated circuits over hybrid circuits did not come until the late 1960s. What is more, until the mid-1970s, magnetic memories made of ferrite cores were the dominant central memory technology, and magnetic hard disks were the dominant storage memory technology.

The invention of the integrated circuit, in February 1959, is attributed to Jack Kilby (Nobel Prize) of Texas Instruments and Robert Noyce of Fairchild Semiconductor.

The development of integrated circuits was based on silicon. Only silicon, thanks to the properties of its oxide, enabled the development of the planar structure of the bipolar transistor, then of the MOSFET transistor and the connections between components incorporated into the substrate.

The take-off of logic and analog integrated circuits in bipolar technology is due to three programs: the Apollo program launched in early 1960, NASA's IMP satellite launched in 1963 and the Polaris sea-to-ground ballistic missile and Minuteman II ground-to-ground missile programs launched in 1962 (section 1.3.4) (CHM (Computer History Museum) 1962).

Integrated circuits based on MOSFET transistors appeared on the market in 1964. In 1971, Intel marketed the first microprocessor, made up of 2,250 MOSFET transistors.

The miniaturization and layout of billions of transistors on a centimetric surface, and thus the realization of microprocessors and memories as the basic building blocks of computers, required technological manufacturing developments in photolithography, etching, layout and connection between components.

This chapter presents:

– the invention of the integrated circuit;

– industrial developments;

– integrated circuit manufacturing technologies.

1.1. Integrated circuits: presentation

The development of integrated circuits took place in two phases:

– From 1961, the production of logic and analog circuits was based on bipolar transistors. Circuits based on MOSFET transistors did not appear until 1970, after a long process of fine-tuning the manufacture of these transistors.

– The development of memories was based on MOSFET transistors in 1970.

This chapter is devoted to logic and analog integrated circuits based on bipolar and MOSFET transistors. Memories are discussed in Chapter 2.

1.1.1. *Logic circuits*

The feasibility of executing arithmetic and logic operations by circuits consisting of electromechanical relays/switches was demonstrated by Claude Shannon in 1937. Commercial "calculators" were built with vacuum tubes (diodes and triodes) until 1954, then with (solid-state) diodes combined with vacuum tubes.

But diodes could only perform AND and OR logic functions. What is more, they could not restore the signal at the output of a logic gate, ruling out the possibility of creating cascades of gates, and hence integrated circuits.

Only the transistor could be used to create universal logic function circuits by combining AND-NO (NAND), and OR-NO (NOR) circuits and restoring the signal at the output of each circuit, thus enabling cascades of logic circuits, and hence creating integrated circuits.

1.1.2. *Analog circuits*

Signal processing circuits include amplification, filtering, modulation and demodulation.

Signal amplification began with the invention of the triode by Lee de Forest in 1906. The invention of the feedback principle by Bell Labs in the early 1930s led to the development of triode-based (vacuum tube) feedback amplifiers in the early 1940s. These were gradually replaced by miniaturized transistor-based circuits.

1.2. The invention of the "integrated circuit"

Overarching all of this, however, at the end of the 1950s, was the emergence of a revolutionary new development, the actual invention and development of the integrated circuit. (Seitz and Einspruch 1998, p. 20)

The invention of the integrated circuit provided the jump-off point for the real electronic revolution. (Orton 2009, p. 98)

1.2.1. *Precursors*

The invention of the integrated circuit, in February 1959, is attributed to Jack Kilby of Texas Instruments and Robert Noyce of Fairchild Semiconductor. Jack Kilby won a Nobel Prize for this work in 2000 (Robert Noyce was ineligible since he had died in 1990).

The concept of the integrated circuit was first formulated by Geoffrey Dummer: an English engineer, in 1952. In September 1957, he presented a circuit: a bistable flip-flop (see note 3) consisting of four transistors, four resistors and seven capacitors (Lojek 2007, p. 37).

In the early 1950s, a number of companies began to investigate the possibility of creating "integrated circuits" based on logic "gates" incorporating the elementary components – resistors, capacitors, diodes and transistors – on a single substrate.

As early as 1949, Jacobi of Siemens filed a patent describing an amplifier ("a multistage transistor amplifier") consisting of five point-contact transistors whose interconnected pairs of points were in contact with a crystalline semiconductor wafer (Jacobi 1952).

In May 1952, Sidney Darlington of Bell Labs patented the design of an amplifier consisting of two bipolar transistors (Si or Ge based) with a common collector, the emitter of one (12A) being connected to the base of the other (13B) by a metal wire (15), as shown in Figure 1.1. The resulting amplification factor is equal to the product of the amplification factors of each transistor (Darlington 1953).

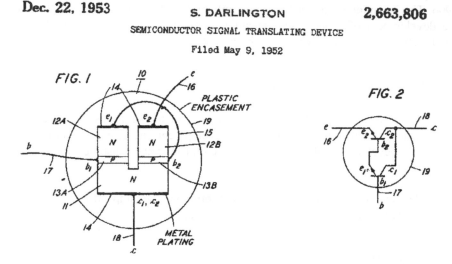

Dec. 22, 1953   S. DARLINGTON   2,663,806

SEMICONDUCTOR SIGNAL TRANSLATING DEVICE

Filed May 9, 1952

Figure 1.1. *Amplifier consisting of two bipolar transistors (Darlington 1953)*

In 1953, Harwick Johnson, of the RCA company, filed a patent, granted in 1957, of a "phase-shift oscillator"[1] (analog circuit), made up of a PN junction (58) constituting an RC circuit (delay line) and a transistor (14), implanted in a germanium substrate (Figure 1.2) (Johnson 1957).

In 1959, Wallmark and Marcus of RCA made several circuits, including a "shift register", an AND (multiple AND) circuit and a half adder, where all passive and active components were integrated into a semiconductor wafer without wire interconnections. But no further development seems to have taken place (Wallmark and Marcus 1959a, 1959b).

---

1 Phase-shift oscillator: a circuit whose function is to produce a periodic signal, sinusoidal, square, sawtooth, etc., generating radio-frequency waves from 20 Hz to 20 kHz.

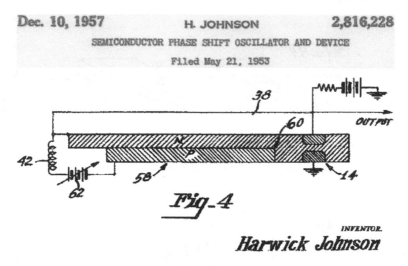

Dec. 10, 1957   H. JOHNSON   2,816,228

SEMICONDUCTOR PHASE SHIFT OSCILLATOR AND DEVICE

Filed May 21, 1953

Figure 1.2. *Phase-shift oscillator (Johnson 1957)*

1.2.2. *Jack Kilby's invention*

Jack Kilby[2] of Texas Instruments made a "phase-shift oscillator" in September 1958, consisting of a germanium mesa bipolar transistor (Volume 1, Chapter 5, section 5.3.2), three resistors and a capacitor, implanted in a germanium wafer and connected by gold flying wires (Figure 1.3).

Jack Kilby, who had just been recruited by Texas Instruments, designed and built this first integrated circuit on his own during a vacation when he was alone in the laboratory.

The management, back from vacation, were thrilled by this achievement:

> The men in the room looked at the sine wave, looked at Kilby, looked at the chip, looked at the sine wave again. Then everybody broke into broad smiles. A new era in electronics had been born (Reid 1985).

---

2 "It suddenly occurred to him that all parts of a circuit, not just the transistor, could be made out of silicon. At the time, nobody was making capacitors or resistors out of semiconductors. If it could be done then the entire circuit could be built out of a single crystal – making it smaller and much easier to produce. [...] Kilbys solution to this problem has come to be called the monolithic idea" (History-Computer n.d.).

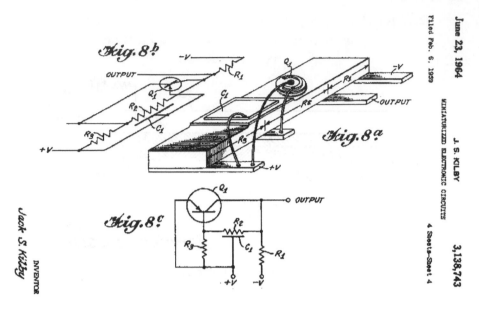

Figure 1.3. *Phase-shift oscillator (Kilby 1964a)*

On February 6, 1959, Kilby filed a patent for the design and manufacture of two integrated circuits (Kilby 1959, 1964a, 1976):

– the phase-shift oscillator implanted in a germanium substrate (Figure 1.3);

– a multivibrator: bistable flip-flop[3] (Figure 1.4) consisting of two mesa bipolar transistors (Volume 1 Chapter 5, section 5.3.2), electrically isolated from each other by air gaps, with four resistors and two capacitors implanted in a semiconductor bar and connected by "flying" gold wires.

The multivibrator circuit designed by Kilby was implemented on a silicon substrate (all-silicon multivibrator circuit) (Lojek 2007, p. 133, Figure 4.28).

Stewart also of Texas Instruments patented a fully integrated NOR circuit on February 12, 1959, consisting of two bipolar transistors (23 and 25), a diode (24) and a resistor (26) (Figure 1.5) (Stewart 1964).

---

3 Multivibrator: bistable flip-flop: basic memory cell. Such a memory component remembers, for as long as it is powered, the last action performed on the R and S buttons (presented in Chapter 2).

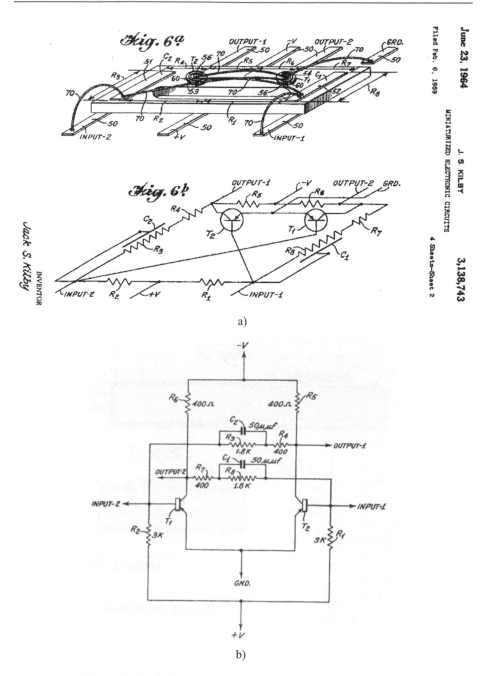

Figure 1.4. *(a) Bistable flip-flop multivibrator (Kilby 1964a);*
(b) corresponding circuit (Kilby 1964b)

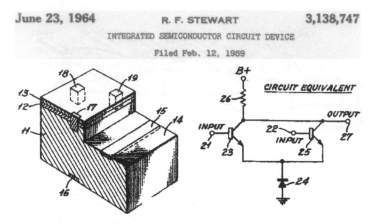

June 23, 1964   R. F. STEWART   3,138,747
INTEGRATED SEMICONDUCTOR CIRCUIT DEVICE
Filed Feb. 12, 1959

Figure 1.5. *OR-NO (NOR) logic integrated circuit (patent figures 2 and 3) (Stewart 1964)*

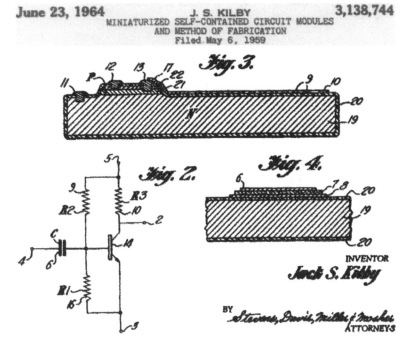

June 23, 1964   J. S. KILBY   3,138,744
MINIATURIZED SELF-CONTAINED CIRCUIT MODULES
AND METHOD OF FABRICATION
Filed May 6, 1959

Figure 1.6. *Kilby's integrated circuit (Kilby 1964b)*

On May 6, 1959, Kilby filed a patent for a circuit (Figure 1.6) consisting of an NPN (14) bipolar transistor (Si or Ge) with a mesa structure (Volume 1, Chapter 5, section 5.3.2) (19: *collector*; 21: *base*; 22: *emitter*), three resistors (R1, R2, R3) and a capacitor C (6: *upper conducting film*; 7: *dielectric film*; 8: *lower capacitor*) connected by metal wires (11, 12, 13) deposited in openings in the germanium substrate (Kilby 1964b).

1.2.3. *Robert Noyce's patent*

In July 1959, Robert Noyce, co-founder of Fairchild Semiconductor in 1957, from a transfer from Shockley Semiconductor Laboratory, was aware of Kilby's patent and asked Jay Last to start an integrated circuit development program. The team decided to build a bistable flip-flop.

Robert Norman designed the circuit, consisting of four transistors (Figure 1.7(A)), while the layout was designed by Lionel Kattner (Figure 1.7(B)). The device is built on a silicon substrate incorporating four planar NPN bipolar transistors (Volume 1, Chapter 5, section 5.3.3), isolated from each other by epoxy resin-filled trenches in the silicon substrate (Figure 1.7(B) (figure 9)) and connected by metal wires (resistors) deposited by the CVD process (Volume 1, Chapter 5, section 5.2.4.1) in openings in the protective oxide layer (Norman et al. 1960). The first circuit was produced in May 1960 (Last 1964; Lojek 2007, p. 136).

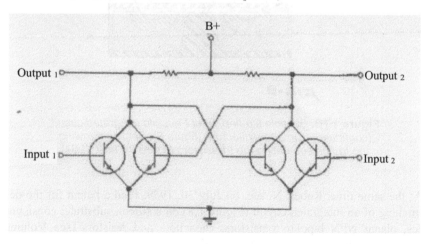

Figure 1.7A. *Bistable flip-flop. First Fairchild integrated circuit (Lojek 2007, p. 134)*

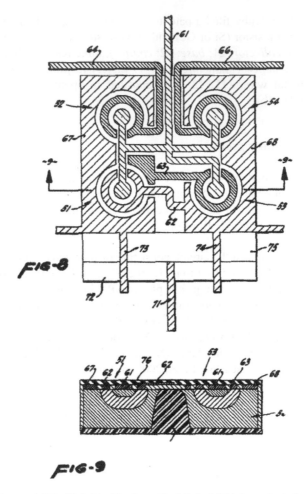

Figure 1.7B. *Bistable flip-flop. First Fairchild integrated circuit (continued). Circuit architecture (figure 8). Section showing the trench separating the transistors (figure 9) (Last 1964)*

At the same time, Robert Noyce, on July 30, 1959, filed a patent for the design and making of an integrated circuit (Figure 1.8) on a silicon substrate, consisting of diodes, planar NPN bipolar transistors, capacitors and resistors (see Volume 1, Chapter 5, section 5.3.3), capacitors and resistors connected by metal wires deposited by CVD process in openings in the protective oxide layer. The main claim is that components are connected by aluminum wires embedded in the substrate (Noyce 1961).

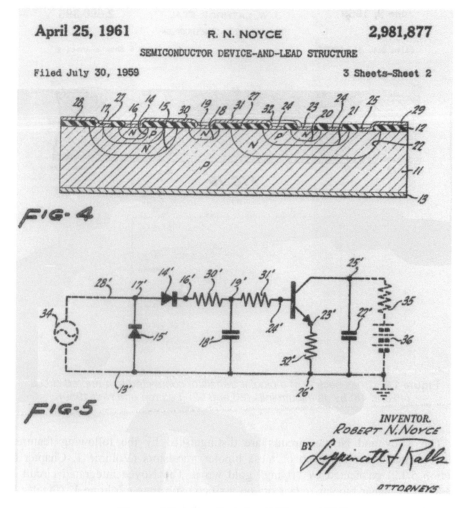

Figure 1.8. *Integrated circuit (Noyce 1961)*

The authorship of this claim is disputed by Bo Lojek (2007, pp. 146–154). In fact, metallic connections (60 and 62) between a transistor and a printed circuit (46 and 48) produced by the CVD aluminum deposition process were described by Jay Lathrop and James Nall of Army Diamond Ordnance Fuze Laboratories (DOFL), in a patent filed in October 1957 (Figure 1.9), of which Robert Noyce was aware of, having visited them (Nall and Lathrop 1958; Lathrop and Nall 1959). Furthermore, in Jack Kilby's patent 3138744 filed on May 6, 1959 (Figure 1.6), interconnections were also produced by CVD deposition of metal wires.

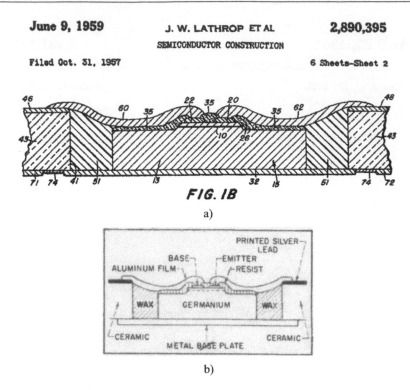

Figure 1.9. *Cross-section of a bipolar transistor connected to a printed circuit (46 and 48) by aluminum foils (60 and 62) (Lathrop and Nall 1959)*

The Kilby and Noyce circuits are distinguished by the following features: Kilby's circuits are made up of mesa bipolar transistors (Volume 1, Chapter 5, section 5.3.2) connected by "flying" gold wires. The Noyce integrated circuit is made up of planar bipolar transistors on a silicon substrate (Volume 1, Chapter 5, section 5.3.3), and the components are connected by metal wires embedded in the substrate (deposited on the substrate using the CVD process). We will see later that Texas Instruments was forced by its financial backers to adopt the planar structure for its transistors, which allowed the connections to be integrated at the base, thus enabling the production of "real" integrated circuits.

Noyce's patent, filed on July 30, 1959 and granted on April 25, 1961, while Kilby's had been filed on February 6, 1959 but granted on June 23, 1964, enabled Noyce to claim paternity for the "invention" of the integrated circuit. Kilby was awarded the Nobel Prize in 2000 (Robert Noyce was ineligible as he had died by this point).

Nevertheless Bob Noyce and Gordon Moore were skeptical ("we are still somewhat skeptical") about the value of the integrated circuit (Fairchild, Progress Report, R&D, January 9, 1961) (Lojek 2007, p. 139).

This attitude led to the resignation of Jay Last and Jean Hoerni, on January 31, 1961, who set up Amelco Semiconductor, a division of Teledyne. Hoerni devoted himself to the development of analog integrated circuits (section 1.3.4).

But since Texas Instruments and Westinghouse were developing and advertising molecular electronics analog integrated circuits for the Minuteman missile program (section 1.3.4), Robert Noyce had no choice but to pursue the development of integrated circuits.

1.2.4. *Lehovec's invention*

The achievement of bipolar transistor integrated circuits became effective with the invention of Kurt Lehovec of the Sprague Electric Company in 1958 on the electrical isolation of bipolar transistors (Lehovec 1962).

In an integrated circuit, the bipolar transistors had to be isolated from each other. To isolate them, they were made in islands: N-islands in a P-substrate (Figure 1.10). Sprague licensed Lehovec's patent to Fairchild and TI.

April 10, 1962   K. LEHOVEC   **3,029,366**

MULTIPLE SEMICONDUCTOR ASSEMBLY

Filed April 22, 1959

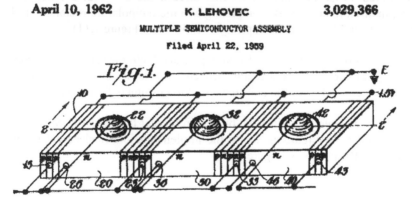

Figure 1.10. *Circuit consisting of a chain of bipolar germanium transistors separated by PN junctions (Lehovec 1962)*

1.3. Industrial developments

1.3.1. *Hybrid circuits*

In the late 1950s, RCA, under contract to the US Army Signal Corps, developed hybrid circuits made up of transistors and passive components mounted on ceramic supports and connected together by wires.

In 1964, IBM developed hybrid circuits of silicon bipolar transistors assembled on ceramic substrates (solid logic technology (SLT)) for its IBM 360/50 and 65 family of computers. IBM produced hundreds of millions of SLT modules in a highly automated factory built for this purpose (Davis et al. 1964).

Bell Labs produced hybrid circuits for their telephone systems until the late 1960s (Lepselter 1966).

The first generation of hybrid circuits in 1964 featured just one logic function per module. In 1967, the second generation featured two to five logic functions per module.

1.3.2. *Logic function integrated circuits*

In March 1960, TI launched its first circuit: the SN 502, a diffused silicon bistable multivibrator network, comprising two mesa bipolar transistors, four diodes, four resistors and four capacitors, priced at $450 each (Figure 1.11).

The transistors were insulated by air gaps etched into the substrate and connected by flying gold wires. The connections between the various circuit components were a major problem. A few dozen units were sent to customers for evaluation[4].

The first integrated circuits were built with bipolar transistors in early 1961. The development of the circuits is due to the implementation of electrical isolation of transistors by PN junctions (Figure 1.10).

---

4 TI advertisement from April 1960 stated that: "this multivibrator, the TI type 502, is so real it carries a price tag; $450 per circuit in quantities less than 100, $300 each for larger quantities" (Lojek 2007, p. 235).

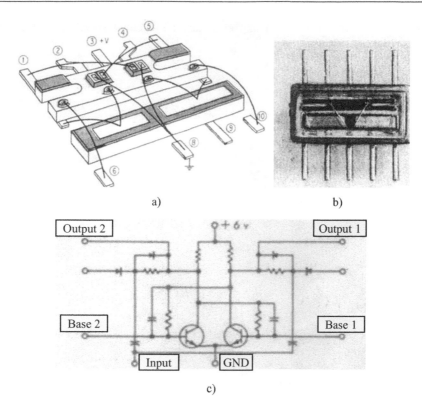

a)   b)

c)

Figure 1.11. *(a) Multivibrator, Texas Instruments type 502; (b) the component; (c) circuit (Lojek 2007, p. 237, Figures 7.7 and 7.8)*

1.3.2.1. *RTL, DTL and TTL logic function circuits*

There have been three generations of logic function circuits based on bipolar transistors: resistor transistor logic (RTL), diode transistor logic (DTL) and transistor transistor logic) (TTL). The three successive generations of NAND circuits are shown in Figure 1.12. The NAND logic function enables all logic functions to be realized.

The first RTL generation was based on discrete-component circuits, with resistors being the cheapest components (Figure 1.12(a)).

The second generation of DTL circuits offered lower power consumption and higher switching speed than RTL circuits (Figure 1.12(b))

The third generation of TTL circuits, derived from the DTL circuit in which the diode group is replaced by transistors or a multiple-emitter ECL transistor, offered even lower power consumption and even faster switching speeds than the DTL circuits (Figure 1.12(c))

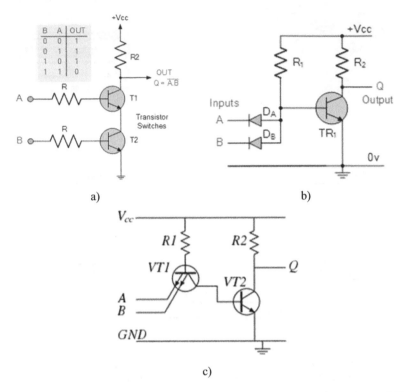

Figure 1.12. *AND-NO (NAND) circuits: (a) two-input RTL; (b) two-input DTL; (c) two-input TTL A and B (Electronics Tutorials n.d.). For a color version of this figure, see www.iste.co.uk/vignes/silicon2.zip*

1.3.2.2. *The first commercial integrated circuits*

Take-off of integrated circuits are due to three programs: the Apollo program launched in early 1960, NASA's IMP satellite launched in 1963 and the Polaris sea-to-ground ballistic missile and Minuteman II ground-to-ground missile programs launched in 1962 (section 1.3.4) (CHM (Computer History Museum) 1962).

The first commercial integrated circuits were built with planar bipolar transistors structure in early 1961.

In 1961, Fairchild launched a number of RTL circuits, designed by Bob Norman (CHM puter History Museum) 1960; Norman et al. 1960; Lojek 2007, p. 145:

1) F-Flip-Flop (four transistors and two resistors): µlogic 923 RTL;

2) S-Half-shift register (nine transistors and five resistors);

3) G-Gate (three transistors and one resistor): µL903 (three-input *NOR*);

4) B-Buffer (three transistors and three resistors);

5) H-Half Adder (four transistors and three resistors);

6) C-Counter Adapter (six transistors and five resistors);

7) R-Shift Register (17 transistors and nine resistors).

Figure 1.13. *Fairchild Micrologic CI µL903 (Lojek 2007, p. 145). For a color version of this figure, see www.iste.co.uk/vignes/silicon2.zip*

The µL903 G-type logic circuit (three-input *NOR*) was the basic component of the Apollo rocket guidance computer (5,000 µL903) and the first major application of integrated circuits (Figure 1.13).

The $25 billion Apollo (man on the moon) program, which began in early 1960 and was launched in 1969, gave a real boost to research into integrated circuits[5]. NASA's Apollo Guidance Computer (ACG), designed by MIT in 1962, manufactured by Raytheon and housed in the Apollo rocket's nose cone, featured 4,100 µL903 NI type G (three-input NOR gate) chips in a first version, and 2,800 NI (dual three-input NOR gate) RTL circuits, manufactured by Fairchild

5 "The decision of President Kennedy in 1961 to mount an intensive space program, with the intention to put a man on the moon in 1970, kick-started a technological revolution, certainly no other country ever received a comparable boost. It became clear that the required rocket guidance would demand highly sophisticated computer technology and that such advanced circuitry could only be realized in integrated form" (Orton 2004, p. 97).

Semiconductor. In July 1969, Armstrong set foot on the Moon. By then, the Apollo program had already purchased over a million of these chips.

Texas Instruments, having adopted the planar structure for its silicon transistors, announced in October 1961 its first RTL circuits with interconnections produced by metal CVD process: the SN 512 NOR/NAND gate; SN 514 dual NOR/NAND; SN 515 exclusive-OR gate; as well as flip-flop bistable circuits 510-511 (Figure 1.14).

Figure 1.14. *Dual NOR/NAND RTL integrated circuit, Texas Instruments (History-Computer n.d.). For a color version of this figure, see www.iste.co.uk/vignes/silicon2.zip*

NASA's IMP (Interplanetary Monitoring Probe Platform) satellite (measuring cosmic rays, solar wind and interplanetary magnetic fields), launched on November 27, 1963, is equipped with integrated circuits manufactured by TI (SN 510 and SN 514), which were the first to orbit the Earth (CHM (Computer History Museum) 1962).

Soon after, other companies began developing integrated circuits.

In 1963, the second generation of DTL circuits were developed by Signetics: Utilogic SE100, a company set up by defectors from Fairchild (CHM (Computer History Museum) 1963).

Fairchild simultaneously developed the DTL 930 series.

In 1963, the third generation appeared: TTL circuits. The first TTL circuits were designed by James Buie of TRW Semi-conductors Inc. in 1961 (Buie 1966).

TTL circuits formed the structures of SRAM memories, themselves made up of flip-flop bistables as early as 1965 (Chapter 2, section 2.2.3).

The first TTL circuits were developed by Sylvania in 1963. Thomas Longo developed three circuits (NAND), (NOR), (bistable flip-flop) (implanting over 500 circuits on a 1" diameter wafer) (Longo et al. 1963) (see section 2.2.3).

Given the success of Sylvania's TTL circuits, Texas Instruments launched the SN 5400 TTL series, a replica of the Sylvania circuits, in 1964. A year later, in 1965, TI launched the SN 7400 series of two-input NAND circuits and the 7474 series of bistable flip-flops.

Although Fairchild had the technology to produce TTL circuits, Fairchild was slow to respond. Finally, Fairchild responded in 1966 by creating integrated circuits with 30–50 gates (9300 series of MSI devices: shift registers, counters, arithmetic logic units) (Laws 2015).

In 1971, TI responded with the 74S series of TTL circuits to replace the SN 7400 series made up of, bipolar transistors incorporating Schottky diodes (Volume 1, Chapter 2), enabling switching speeds of 3 ns per gate. TI remained the dominant manufacturer.

Eleven years after creating a gigantic market for cheap transistors by launching pocket radios, TI, in 1967, launched pocket calculators using the (NOR) circuit.

The first MOSFET transistor-based integrated circuit appeared in 1962. It was with these circuits that memories and microprocessors were created.

In 1962, the first experimental NOR circuit consisting of 16 MOSFET transistors was produced by Steve Hofstein and Fred Herman from RCA (Laws 2011), and in 1964 the first commercial IC with 120 PMOSFET transistors (a 20-bit shift register) by GME (Norman and Stephenson 1969).

In 1968, RCA launched the 4000 series of logic integrated circuits based on CMOS components (Medwin 1968, 1971). In 1968, Federico Faggin at Fairchild-SGS with Thomas Klein produced the first integrated circuit based on self-aligned silicon gate MOSFET transistors (see Volume 1, section 6.2.2.1): the Fairchild 3708 (Faggin et al. 1969).

1.3.3. Microprocessors

The microprocessor, or programmable chip (MPU), is the processing unit in a computer: integrated circuit made up of a set of integrated logic circuits and memories (registers). Only microprocessors based on MOSFET transistors have ever been produced.

In 1971, Intel marketed the first microprocessor, Intel 4004, consisting of 2,250 silicon gate MOSFET (PMOS) transistors (Volume 1, Chapter 6, section 6.2.2.1), with a channel length of (line width) 10 μm, designed and built by Martin Hoff and Federico Faggin (Faggin and Hoff 1972) (Faggin coming from Fairchild). In 1974, Intel marketed the 8080 microprocessor (4,500/6,000 transistors), which is considered to be the first true microprocessor (CHM (Computer History Museum) 1971).

At the same time, the following microprocessors were developed: the MOTOROLA 68000 in 1974, which was to be the microprocessor for the Apple Macintosh; the Intel 8088 microprocessor for the IBM PC; and the Zilog Z80 microprocessor. It should be noted that the designer of the Zilog microprocessor was Federico Faggin, who left Intel to set up Zilog, followed by:

– in 1985: Intel 386, a microprocessor with 275,000 transistors and a gate width of 1.5 μm;

– in 1989: Intel 860, a microprocessor with 1 million transistors operating at 32 Gflops (32×10^9 operations/s);

– 1993: Intel Pentium, processor with 3.1×10^6 transistors (0.8 μm biCMOS process), speed 66 MHz;

– in 2002: Intel Pentium 4: 42 million transistors, speed 1.2 GHz;

– in 2015: the sixth generation of Intel Core processors (with a 14 nm etch) was launched.

Microprocessor development became essentially a design problem, as Gérard Berry points out: "the design of such circuits is essentially very high-end software". This ushered in a new era: that of the software engineer. Today's chips can only be designed with the help of sophisticated CAD programs to manage their complexity (Berry 2017, p. 88).

1.3.4. Analog circuits

Circuits for processing weak signals include amplification, filtering, modulation and demodulation.

The first hybrid "operational amplifier"[6], using bipolar silicon transistors (instead of triodes), was designed and built by Georges Philbrick of Philbrick Research (GAP/R model P65) and produced from 1961 to 1971. It consisted of three conventional stages, each with two bipolar transistors (Figure 1.15) (Jung 2006).

---

6 Analog calculators were the first application of op-amp amplifiers. An op-amp can perform the mathematical operations of calculation and was called an operational amplifier.

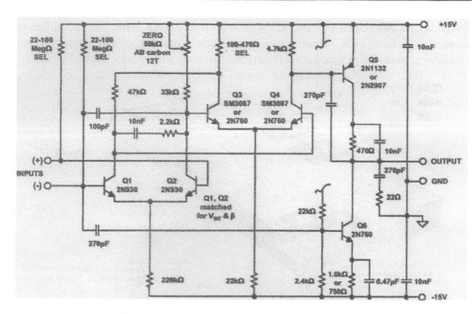

Figure 1.15. *First "hybrid" operational amplifier based on "discrete" bipolar silicon transistors, the GAP/R model P65 op amp (Jung 2006)*

The most common architecture of today's operational amplifier consists of three stages in series: a differential stage, a gain and offset stage and a power stage (Figures 1.15 and 1.19).

The differential stage is based on a pair of identical transistors (bipolar or field-effect) connected symmetrically.

The structure of the gain and offset stage varies from one technology to another. Its role is to provide the high gain (120 dB typical) and to supply the ad hoc electrical levels so that the output amplifier stage is at the mid-point of the power supply at rest.

The output stage features a pair of complementary transistors in push-pull configuration, that is, one conducts when the other is not. This structure results in low output impedances. This enables the amplifier to deliver high output currents with low output impedance.

In 1962, Texas Instruments and Westinghouse developed the first analog integrated circuits for the Autonetics guidance system for Minuteman II ground-to-ground missiles (ICBMs). One version of the differential amplifier designed and developed by Jack Kilby is shown in Figure 1.16. This amplifier is made up of four NPN silicon mesa transistors. Note the connections between transistors: "with loose wire interconnects, surprisingly the circuits worked".

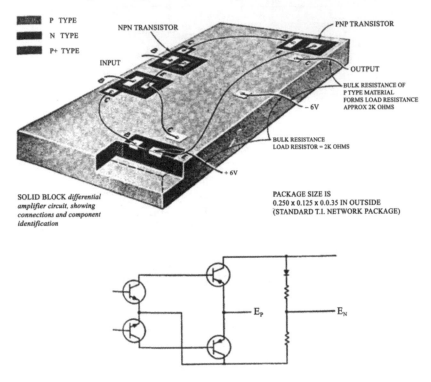

Figure 1.16. *Differential amplifier built in a monolithic silicon P block by Jack Kilby (Lojek 2007, p. 236)*

At the same time, H.C. Lin at Westinghouse designed a differential amplifier for this program (Figure 1.17).

In 1962, Autonetics, responsible for the development of the Minuteman missile guidance system, forced Westinghouse and Texas Instruments to abandon their technology, based on the mesa transistor and flying-wire connections between the components, for a technology developed by Fairchild, that is, an integrated circuit whose basic component is a planar transistor. The resulting circuits were used in the missile guidance system for the Minuteman II (ICBM): TI (18 circuits), Westinghouse (18 circuits), RCA (1 circuit), GE (1 circuit) (Lojek 2007, p. 156). This program became the biggest consumer of integrated circuits.

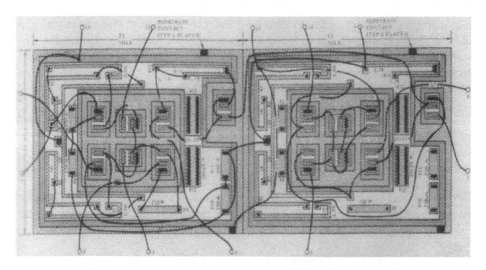

Figure 1.17. *Prototype of a differential amplifier with flying leads between components (Lojek 2007, p. 243)*

Fairchild was also involved in the development of such amplifier circuits through "defense" contracts in early 1963. A differential amplifier integrated circuit µA001 (5T+7R+3D)" was produced. But manufacturing this amplifier ran into problems.

In 1963, Robert Widlar, recently recruited by Fairchild, designed and built what was to become the first commercial operational amplifier integrated circuit µA702, consisting of a nine planar silicon NPN bipolar transistor structure and 11 diffusion resistors (Figure 1.18) (Lojek 2007, p. 269).

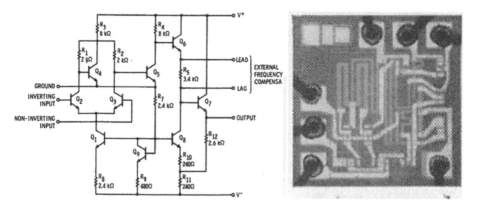

Figure 1.18. *Operational amplifier circuit µA702 designed by Robert Widlar of the Fairchild company (Lojek 2007, p. 269)*

According to Bo Lojek, Widlard and Talbert carried out this circuit almost clandestinely, finding an ally in Floyd Kvamme from the marketing department. The response from customers was overwhelmingly positive, and it was then that Robert Noyce took an interest in Bob Widlar and his µA702 integrated circuit and decided to market it immediately in October 1964 (Lojek 2007, p. 270).

The µA702 circuit was followed by the µA709 (14T+15R) circuit in 1965, and also created by Bob Widlar, one of the semiconductor industry's greatest commercial and technical successes (Widlar 1966). Bendix ordered 10,000 units in the December.

Hoerni and Last had left Fairchild in January 1961 to create the Amelco company, which became a major supplier of linear circuits: the Amelco 809BE model, a competitor to the µA709 model (Lojek 2007, p. 277).

Talbert and Widlar left Fairchild at the end of 1965 for Molectro (later acquired by National Semiconductor), where they built an analog circuit line with the LM101 model (Widlar 1967).

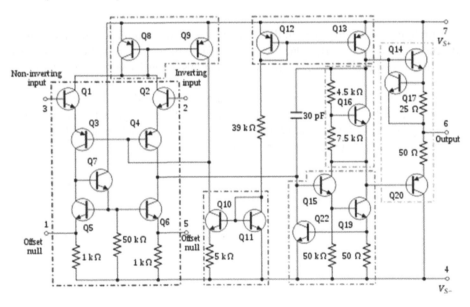

Figure 1.19. *Fairchild Op-Amp 741 operational amplifier (20T, 11R, 1C).* ■*: Blue denotes the differential input stage;* ■*: red denotes the current mirrors;* ■*: cyan denotes the output stage;* ■*: magenta denotes the voltage amplification stage;* ■*: green denotes the output stage biasing device (Wikipedia 2023). For a color version of this figure, see www.iste.co.uk/vignes/silicon2.zip*

In 1968, Dave Fullagar of Fairchild designed and built the µA741 model, which offered better performance, while being more stable and simpler to implement (Figure 1.19).

Although its performance was similar to that of its main competitor, National Semiconductor's LM101, the µA741 became a standard because it was easier to use than the LM101. The µA741 is still manufactured today (CHM (Computer History Museum) 1964).

By 1966, amplifier ICs were being produced by a number of companies: Amelco, General Electric, Motorola, RCA, TI and General Instrument.

1.4. Integrated circuit technologies

1.4.1. Integrated circuit design and photomask production

Circuit design is the layout of all circuit components and their interconnections, and the making of photomasks preceding manufacturing operations.

An integrated circuit is made up of a stack of layers. Each layer is a juxtaposition of zones with a specific function. The function is performed by a specific process.

The growing complexity of integrated circuits has necessitated a parallel development of circuit design tools. ICs can only be designed with the help of sophisticated CAD programs that manage their complexity. Once the design is complete, the computer's memory contains the positional coordinates, lateral and in depth dimensions of each component part. These data are used to draw up the plans for each layer and, per layer, the plans for each family of zones corresponding to a manufacturing operation. The plan of each family of zones corresponding to an operation is then generated automatically on a photographic plate, to scale 10, by scanning a light beam. The photomasks are produced by optically projecting the plan onto a photographic plate.

1.4.2. The manufacturing process

The manufacturing of an integrated circuit involves the following successive operations, some of which are described in Chapters 5 and 6 of Volume 1:

– manufacture of n- or p-doped single-crystal silicon wafers (Volume 1, Chapter 4, section 4.2);

– deposition of a lightly doped epitaxial layer (Volume 1, Chapter 5, section 5.3.4; section 1.4.3.1);

– oxidation masking: making of an oxide layer, constituting a selective mask (Volume 1, section 5.3.1; section 1.4.3.2);

– photolithography process by which a pattern (design, plane) is transferred from a photomask onto a layer of sensitive photoresist material (see section 1.4.3.3)

– etching: making openings in an N or P layer or in the oxide layer, chemical etching or ion etching (see section 1.4.3.4);

– doping by diffusion (Volume 1, section 5.2.4) or ion implantation (Volume 1, section 5.2.5);

– deposition in apertures of a layer by physical vapor deposition (PVD) or chemical vapor deposition (CVD) (Volume 1, section 5.2.4.1): making connection wires between components.

Figure 5.20 (Chapter 5) in Volume 1 shows the steps involved in building a bipolar NPN transistor with a planar structure, and Figure 6.9 (Chapter 6) in Volume 1 shows the steps involved in building a MOSFET transistor.

The main steps (Figure 1.20) are as follows: step 1: oxidation of the silicon wafer with formation of an oxide layer; step 2: application of a photosensitive emulsion (photoresist); step 3: exposure of the photoresist layer through a mask to an ultraviolet light beam; step 4: chemical dissolution of the exposed regions of the photoresist layer; step 5: etching of the oxide layer; step 6: removal of the photoresist; step 7 (not shown): making up, by diffusion or ion implantation through the windows of the oxide layer, of the source and drain of a MOSFET transistor.

This process is repeated several times with different masks for integrated circuits.

Miniaturization has necessitated the development of specific technologies for creating windows in the oxide layer: photolithography and etching.

1.4.3. *The technologies*

1.4.3.1. *The epitaxial layer*

For both types of transistor, the deposition of an epitaxial surface layer of lightly doped silicon on the monocrystalline substrate became an essential step in the integrated circuit production process (Volume 1, Chapter 5, section 5.3.4).

When building integrated circuits with NPN or PNP bipolar transistors, the problem is in isolating each transistor (Figures 1.8 and 1.10). To isolate the transistors from each other, they are formed inside islands: N islands in a P substrate. The silicon N highly-doped wafers are covered with a silicon epitaxial layer of very high resistance (pure) silicon. A low-doped n layer is diffused back into the epitaxial layer. The intermediate p zones in the epitaxial layer are created by boron diffusion through photolithographically defined openings. PN junctions are thus formed, constituting diodes connected in reverse polarity (Figure 1.10). Fairchild and Texas Instruments adopted this technology for their integrated circuits based on planar bipolar transistors (see Volume 1, Chapter 5, section 5.3.4).

For a MOSFET transistor, an undoped epitaxial layer deposited on the substrate allows local implantation, by means of masks, of a low-doped p "inversion channel" region, thereby controlling the threshold voltage (Volume 1, Chapter 6, section 6.2.2.1).

1.4.3.2. *Oxide masking and patterning*

It was Carl Frosch and Lincoln Derick's discovery of the exceptional protective properties of the silica oxide layer produced in situ by oxidation treatment on the silicon substrate of transistors that enabled photolithography to produce bipolar transistors, then MOSFETs with a planar structure. Derick and Frosch then showed that the oxide layer could be used as a selective mask to define doping zones (emitter and receiver of a bipolar transistor, or source and drain of a MOSFET transistor) on the substrate by showing that certain dopants could not diffuse through the oxide layer (the latter then constituting a barrier to their penetration into the substrate) and thus that by etching windows in the oxide layer certain well-defined zones could be doped (see Volume 1, Chapter 5, section 5.3.1).

According to Holonyak (2007), "In 1955 this discovery of the protective silicon-dioxide layer, oxide masking and patterning, ultimately led to the integrated circuit".

1.4.3.3. *Photolithography*

Photolithography is in many ways the key to microelectronics technology. (Oldham 1977)

Transistors are made by photolithography. (Nall and Lathrop 1958)

Photolithography: a process by which a pattern (design, plane) is transferred from a photomask by a beam of (ultraviolet) light (step 3) onto a layer of material sensitive (photoresist) to light energy.

The first use of photolithography to produce a bipolar transistor with a mesa structure was by James Nall and Jay Lathrop of the U.S. Army DOFL laboratories in 1957. The patent describes the various steps involved in producing the successive layers of a PNP bipolar transistor, including deposition of a photoresist layer and exposure to ultraviolet light through a mask (Lathrop and Nall 1959).

Advances in lithography are driving progress in miniaturization and therefore the performance of integrated circuits.

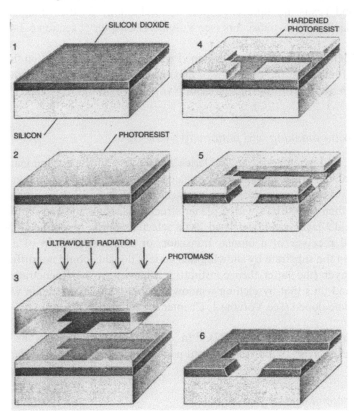

Figure 1.20. *Photolithography (Oldham 1977)*

The photoresist (step 2, Figure 1.20) is a resin sensitive to ultraviolet light, which produces a more or less advanced polymerization of the affected areas, which determines its solubility in certain solvents (step 3) and thus the creation of openings with well-defined edges in the photoresist layer (step 4). Conversely, the remaining areas of the photoresist layer unaffected by the light must withstand the etching of the oxide layer (step 5), which exposes the areas of the semiconductor (step 6) that are to undergo a doping operation, for example.

The quality of the apertures (clean edge) in the successive photoresist and oxide layers is linked to the phenomenon of light diffraction induced by the edges of the photomask apertures. Depending on the wavelength and the distance between the photomask and the photoresist layer, the edges become increasingly blurred. The lower the wavelength of the light beam, the better the resolution (sharper edges).

Miniaturization, which has seen channel length (gate length) decrease from around ten microns in 1970 to around twenty nanometers in 2010[7], demands ever-higher resolution and precision in the superposition (alignment) of successive masks. Hence the use of UV beams with wavelengths from 193 nm (0.20 μm) down to a wavelength of 13.5 nm, the limit of X-ray lithography.

1.4.3.4. *Selective etching*

Figure 1.21 shows the two etching processes: wet chemical etching (isotropic etching) and dry ion etching (free radical etching).

1.4.3.4.1. Wet chemical etching

It was also thanks to a specific property of silica that it was possible to create apertures in the protective oxide layer (step 6), without etching either the photoresist (defining the apertures) or the underlying single-crystal silicon: "the specificity of hydrofluoric acid etching of silica".

Likewise, electrical connections between components using aluminum wires, after deposition of aluminum layers by the CVD process, are made by specific attack with phosphoric acid on the aluminum layer.

1.4.3.4.2. Dry plasma etching (reactive ions)

Ion etching produces a (sputtering) ejection of atoms. This is made possible by the specificity of the plasma gas, depending on the material. For silica: CF_4 or C_2F_6-CHF_3; for silicon nitride: $CF_4 + O_2$; for aluminum: CCl_4-Cl_2; for polysilicon C_2F_6-CF_3Cl. The attack is directional.

Regarding the free-radical etching of silicon (gaseous SF_6 molecules excited by a low-pressure, high-frequency plasma) by adsorption-reaction-desorption processes produces SiF_4 moles, etching is isotropic.

---

7 See Table 6.1 (Volume 1, Chapter 6) summarizing developments in MOSFET transistors.

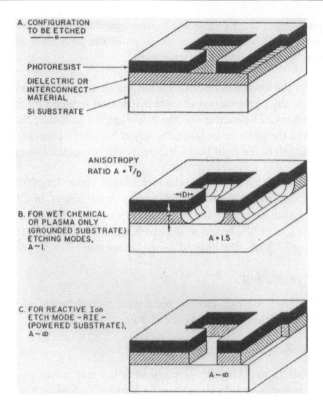

Figure 1.21. *The etching processes (Oldham 1977)*

1.5. References

Berry, G. (2017). *L'Hyperpuissance de l'informatique*. Odile Jacob, Paris.

Brock, D.C. and Laws, D.A. (2012). The early history of microcircuitry: An overview. *IEEE Annals of the History of Computing*, 34(1), 7–19.

Buie, J.J. (1966). Coupling transistor logic and other circuits. Patent, US3283170.

CHM (Computer History Museum) (1960). The Silicon Engine Timeline 1960: First planar integrated circuit is fabricated [Online]. Available at: http://www.computerhistory.org/semiconductor/timeline.html.

CHM (Computer History Museum) (1962). The Silicon Engine Timeline 1962: Aerospace systems are first. Applications for ICS in computers [Online]. Available at: http://www.computerhistory.org/semiconductor/timeline.html.

CHM (Computer History Museum) (1963). The Silicon Engine Timeline 1963: Standard logic IC families introduced [Online]. Available at: http://www.computerhistory.org/semiconductor/timeline.html.

CHM (Computer History Museum) (1964). The Silicon Engine Timeline 1964: The first widely-used analog integrated circuit is introduced [Online]. Available at: http://www.computerhistory.org/semiconductor/timeline.html.

CHM (Computer History Museum) (1971). The Silicon Engine Timeline 1971: Microprocessor integrates CPU function into a single chip [Online]. Available at: http://www.computerhistory.org/semiconductor/timeline.html.

Darlington, S. (1953). Semiconductor signal translating device. Patent, US2663806.

Davis, E.M., Harding, W.E., Schwartz, R.S., Corning, J.J. (1964). Solid logic technology: Versatile, high-performance microelectronics. *IBM Journal of Research and Development*, 8(2), 102–114.

Electronics Tutorials (n.d.). Digital logic gates. Logic NAND gate tutorials [Online]. Available at: electronics-tutorials.ws.

Faggin, F. and Hoff, M.E. (1972). Standard parts and custom design merge in four-chip processor kit. *Electronics*, 112–116.

Faggin, F., Klein, T., Vadasz, L. (1969). Insulated gate field effect transistor integrated circuits with silicon gates. *IEEE Transactions on Electron Devices*, 16(2), 236.

History-Computer (n.d.). History of computers and computing. Birth of the modern computer. The IC history [Online]. Available at: history-computer.com.

Holonyak, N. (2007). The origins of diffused-silicon technology at Bell Labs, 1954–1955. *The Electrochemical Society Interface, Fall*, 30–34.

Jacobi, W. (1952). Halbleiterverstârker (semiconductor amplifier). Patent, 833366(DPMA).

Johnson, H. (1957). Semiconductor phase shift oscillator and device. Patent, US2816228.

Jung, W. (2006). *Op-Amp Application Handbook*. Elsevier, Amsterdam.

Kilby, J. (1959). Semiconductor solid circuits. *Electronics*, 32(8), 110–111.

Kilby, J. (1964a). Miniaturized electronic circuits. Patent, US3138743.

Kilby, J. (1964b). Miniaturized self-contained circuit modules and method of fabrication. Patent, US3138744.

Kilby, J. (1976). Invention of the integrated circuit. *IEEE Transactions on Electronic Devices*, 23, 648–655.

Last, J.T. (1964). Solid-state circuitry having discrete regions of semiconductor material isolated by an insulating material. Patent, US3158788.

Lathrop, J.W. and Nall, J.R. (1959). Transistors are made by lithography. Patent, US2890395.

Laws, D. (2011). Oral history of Steven. R. Hofstein. Interview, Computer History Museum, Mountain View, California.

Laws, D. (2014). Who invented the IC. Interview, Computer History Museum, Mountain View, California.

Laws, D. (2015). The rise of TTL: How Fairchild won a battle but lost the war. Interview, Computer History Museum, Mountain View, California.

Lehovec, K. (1962). Multiple semiconductor assembly. Patent, US3029366.

Lepselter, M.P. (1966). Beam-lead technology. *Bell System Technical Journal*, 45(2), 233–253.

Lojek, B. (2007). *History of Semiconductor Engineering*. Springer, Berlin.

Longo, T.A., Feinberg, I., Bohn, R. (1963). Universal high level logic monolithic circuits. In *1963 International Electron Devices Meeting*, Washington, DC. doi: 10.1109/IEDM.1963.187394.

Medwin, A. (1968). Semiconductor translating circuit. Patent, US3390314.

Medwin, A. (1971). Integrated circuit. Patent, US3588635.

Nall, J.R. and Lathrop, J.W. (1958). Transistors are made by lithography. *Electronics Engineering Edition*, 2, 142–143.

Norman, N.H. and Stephenson, H.E. (1969). Shift register employing insulated gate field effect transistors. Patent, US3454785.

Noyce, R. (1961). Semiconductor device-and-lead structure. Patent, US2981877.

Oldham, W.C. (1977). The fabrication of microelectronic circuits. *Scientific American*, 237(3), 110–126.

Orton, J.W. (2004). *The Story of Semiconductors*. Oxford University Press, Oxford.

Orton, J.W. (2009). *Semiconductors and the Information Revolution*. Elsevier, Amsterdam.

Reid, T.R. (1985). *The Chip: How Two American Invented the Microchip and Launched a Revolution*. Penguin Random House, New York.

Seitz, F. and Einspruch, N.G. (1998). *The Tangled History of Silicon*. University of Illinois Press, Champaign.

Siemens (n.d.). ABC de la microélectronique. A19100-F1-AS-VI-7700.

Stewart, R.F. (1964). Integrated semiconductor circuit device. Patent, US3138747.

Wallmark, J.T. and Marcus, S.O. (1959a). Integrated semiconductor devices. *RCA Enginneer*, 5(1), 42–45.

Wallmark, J.T. and Marcus, S.O. (1959b). Semiconductor devices for microminiaturization. *Electronics*, 32(6), 35–37.

Widlar, R.J. (1966). A monolithic operational amplifier. *Fairchild Semiconductor Application Bulletin.*

Widlar, R.J. (1967). Monolithic Op Amp with simplified frequency compensation. Report, Texas Instrument.

Wikipedia (2023). Transistor-transistor logic [Online]. Available at: https://en.wikipedia.org/wiki/Transistor–transistor_logic.

2

Memories

One major consequence of the invention of the integrated
circuit was a significant revolution in data storage or memory.
(Seitz and Einspruch 1998, p. 214)

The first transistorized memories were "bistable flip-flops". It is Claude Shannon who showed that the bistable flip-flop could store a 0 or 1 bit (Shannon 1948). The bistable flip-flop of bipolar technology is the component of the cache and main memory of some computers from 1965 onwards. By 1970, MOSFET-based memories were replacing them.

For main memories, the single-transistor DRAM cell designed and built by Robert Dennard of IBM in 1968 is, according to Chih-Tang Sahn, the invention comparable to the invention of the transistor itself: "A worldwide consensus". This cell replaced magnetic memories with ferrite cores in the mid-1970s (Sah 1988, p. 1283).

For storage memories, the design and manufacture of the floating-gate MOSFET transistor in 1967, by Kahng and Sze of Bell Labs, led to the replacement of (magnetic) hard disks in the mid-1970s and the development of flash memories (Kahng and Sze 1967; Kahng 1970).

2.1. Introduction

A computer consists of:

– a central processing unit (CPU), comprising an arithmetic-logic unit (ALU) and a control unit;

– memories: cache memory, intermediate memory between the CPU and main memory, high-speed memory, main memory and storage memory.

Until the mid-1970s, magnetic memories with ferrite cores were the dominant core memory technology.

Similarly, until the early 1970s, hard disks (magnetic or electromechanical) were the dominant storage technology.

These memories consist of arrays of cells storing a 0 or 1 bit. The information can be stored in or retrieved from any cell in the matrix (Figures 2.4, 2.8 and 2.12). Memory access, both read and write, is called random access memory (RAM), as opposed to magnetic memory, where access is sequential.

We distinguish:

– SRAM and DRAM (RAM: random access memory) memories: volatile memories, as they do not retain information when the power is cut off;

– ROM (read-only memory) and its derivatives: nonvolatile memories;

– cache memory is a static SRAM memory;

– main memory is a dynamic DRAM memory;

– storage memory is a ROM memory.

2.2. SRAM memories

2.2.1. *Cells and their operation*

SRAM memories are volatile, they do not retain information when the power is cut, but they do not need to be refreshed as long as no action is taken.

The memory function in its most elementary form is made up of cross-coupled inverters (Figure 2.1). The output of each inverter is connected to the input of the other inverter. By looping the two inverter, a logic value can be stored.

A logic 0 at the input of the first inverter turns into a 1 at its output and it is fed into the second inverter which transforms that logic 1 back to a logic 0 feeding back the same value to the input of the first inverter. This creates a stable state that does not change over time. This circuit can store two possible states:

$$Q = 0 \text{ and } Q^* = 1 \text{ or } Q = 1 \text{ and } Q^* = 0$$

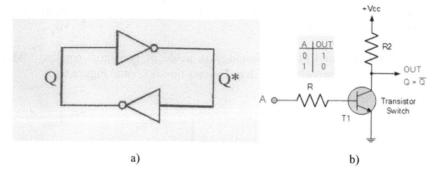

a)   b)

Figure 2.1. *(a) The elementary memory SRAM (Douillard et al. 2006, p. 7);*
(b) the elementary inverter in bipolar logic. For a color version
of this figure, see www.iste.co.uk/vignes/silicon2.zip

The structure of an SRAM cell is shown in Figure 2.2. The basic circuit is a bistable flip-flop with read and write accesses. Writing to and reading from the bit line BL are permitted when the line WL, on which the memory cell is located, is selected.

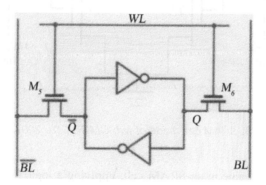

Figure 2.2. *General structure of an SRAM cell (Douillard et al. 2006, p. 7)*

They are the elementary cells of cache memories[1], shift registers and certain main memories, due to their high switching speed and low power consumption.

---

1 The transmission of information between processor (ALU) and main memory is often much slower than the microprocessor's speed potential. To alleviate this problem, an ultra-fast memory is interposed (or integrated) between processor and RAM: a cache memory where the most frequently recurring instructions and data are stored.

SRAM memories based on CMOS components have supplanted memories based on bipolar transistors (Figure 2.3).

The SRAM cell in CMOS technology is made of two cross-coupled CMOS (T1/T3 and T2/T4) (see Volume 1, Chapter 6, section 6.1.4 and Figure 6.6).

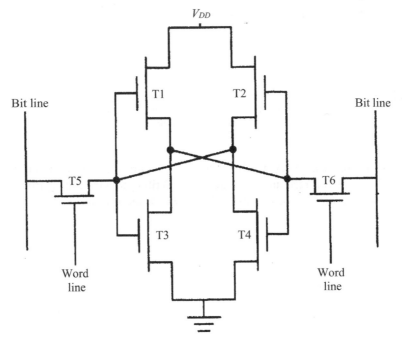

Figure 2.3. *SRAM cell made of two CMOS (Sze 2002, p. 215)*

To write a logic value in an SRAM cell, applying a logic 0 to the Bit Line T5 (data in and out) writes a 0 to the cell, and applying a 0 to the BL line T6 writes a 1. For reading, via the word line (read and write), the BL T5 and BL T6 are connected to the input of a bidirectional amplifier that delivers a logic 1 if B = 1 (BL (T5)) and B = 0 (BL (T6)) and a logic 0 otherwise (see bottom of Figure 2.4).

The first SRAM consisting of six cells of 64-bit PMOSFET cells was produced in 1965 by John Schmidt of Fairchild. In 1968, Fairchild produced a 1,024-bit SRAM (multi-chip memory) by assembling 16/64-bit cells on a ceramic substrate.

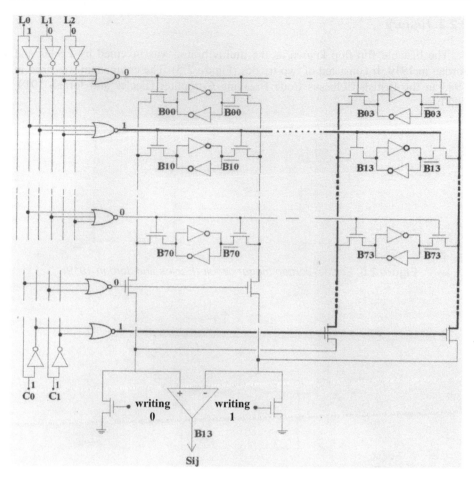

Figure 2.4. *Structure of an SRAM memory consisting of 32 1-bit cells (Douillard et al. 2006, p. 53)*

The structure of an SRAM memory with 32 1-bit (cells) words is shown in Figure 2.4. Address decoding (5 bit) is broken down into 3-bit row addressing (L0, L1, L2) and 2-bit column addressing (C0, C1). For each address, a row and a column are activated and the corresponding MOS switches are closed. For example, when applying address L2L1L0C1C0 = 00111, the second row and fourth column are activated. Writing a 0 or 1 to the cell at the intersection is achieved by applying a logic 0 to the gate of the appropriate transistor.

2.2.2. *History*

The bistable flip-flop known as the multivibrator was invented by Eccles and Jordan in 1919. It consisted of two triodes (Figure 2.5). The component was used in 1943 in the British Colossus Code Breaking Computer (Eccles and Jordan 1919, 1920).

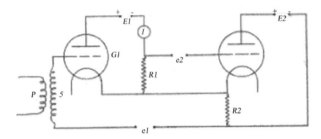

Figure 2.5. *Eccles–Jordan trigger circuit (Eccles and Jordan 1919)*

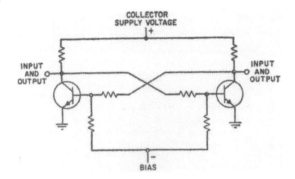

Figure 2.6. *Simple Eccles–Jordan flip-flop (two transistors) (Harris 1958)*

It was Claude Shannon[2] who showed that the flip-flop circuit could store a 0 or 1 bit (Shannon 1948).

---

2 "Early flip-flops were known variously as trigger circuits or multivibrators. A flip-flop circuit has two stable states and, as Claude Shannon pointed out in his "Mathematical Theory of Communica-tion" (1948), a flip-flop can be used to store one bit of information. Flip-flop circuits operate using Boolean algebra (AND, OR, NOT). Thus, with the invention of electronic computing using vacuum tubes as switches, flip-flops became the basic storage element in sequential logic used in digital circuitry, and the basis for electronic memory".

A multivibrator was produced in 1959 by Jack Kilby with two bipolar germanium transistors (Chapter 1, Figure 1.4), presented in his patent 3138743. It won him a Nobel Prize.

In March 1960, TI launched its first SN 502 circuit: a multivibrator consisting of two mesa bipolar silicon transistors (Volume 1, Chapter 5, section 5.3.2; Volume 2, Chapter 1, Figure 1.11).

In 1960, Robert Norman and his colleagues at Fairchild, at the request of Robert Noyce (the other inventor of the integrated circuit), designed and built a bistable flip-flop, consisting of four planar bipolar silicon transistors (Volume 1, Chapter 5, section 5.3.3; Volume 2, Chapter 1, Figures 1.7(a) and (b)); in 1963, Robert Norman patented a solid-state switching and memory apparatus consisting of a matrix of four bipolar transistor flip-flop cells (Figure 2.7) (Norman 1971).

Gordon Moore[3] considered this "idea" "economically so ridiculous" (Moore 1995). Robert Norman resigned from Fairchild in June 1963 and founded General Microelectronics with three Fairchild colleagues.

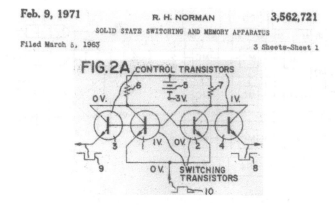

Figure 2.7. *Flip-flop bistable cell (Norman 1971)*

2.2.3. *Industrial developments*

The bistable flip-flop was the basic component of cache and main memory in many computers.

---

3 According to Gordon Moore in a 1995 interview: "In the early days of the integrated circuit Bob Norman suggested the idea of semiconductor memory [...] the whole idea of how semiconductor flip-flops could be used as a memory structure. I decided it was so economically ridiculous, it didn't make any sense to file a patent on it" (Moore 1995).

In 1966, Thomas Longo who had joined the Transitron company, manufactured the first SRAM TMC3162E 16-bit memory, comprising 16 cells forming a 4 × 4 matrix of flip-flops with two bipolar transistors, for the Honeywell model 4200 minicomputer (CHM (Computer History Museum) 1966). Thomas Longo joined Fairchild in 1970.

A cache memory SRAM 16 bit cache memory designed and built in 1965 by Arnold Faber and Eugène Schlig of IBM company, consisting of 64 flip-flop bistable cells (five bipolar transistors and five resistors per memory cell) (Figure 2.8), was used as cache memory on the IBM 360/95 computer in 1968, "the first commercially available computer with cache memory". The main memory was made up of ferrite cores (Faber and Schlig 1967).

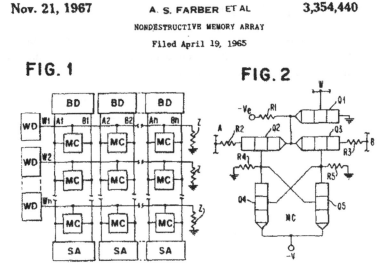

Nov. 21, 1967   A. S. FARBER ET AL   **3,354,440**

NONDESTRUCTIVE MEMORY ARRAY

Filed April 19, 1965

Figure 2.8. *SRAM memory: elementary cell (figure 2) and matrix of elementary cells (figure 1) (Faber and Schlig 1967)*

In 1969, IBM produced a 128-bit SRAM memory using TTL bipolar technology (Chapter 1, section 1.3.2), which was to be used as the main memory for the IBM 370 model 145 computer, released in 1971. In 1973, the same model was equipped with a 1024-bit SRAM memory (Pugh 1981).

Under the direction of Longo in 1970, Fairchild produced a 256 bit TTL SRAM for the main memory of the Burroughs Illiac IV computer.

In 1971, Longo, still at Fairchild, developed a 1 k SRAM (1024-bit RAM). 65,000 chips of this type were used in the central memory of the Cray 1 computer (CHM (Computer History Museum) 1969).

Most of the Fairchild people involved in these developments left Fairchild to join Intel that was founded in 1968. The first integrated circuit produced by Intel in 1969 was a TTL SRAM TTL memory (i3101 64 bit), whose switching speed was twice that of conventional SRAM memories.

By 1970, SRAM memories, based on MOSFET transistors (Figure 2.3), were set to supplant bipolar transistor-based memories.

One of the first commercial components produced by Intel in 1969 was an SRAM 1101, 256bit memory (PMOSFET Si gate: see Volume 1, Chapter 6, section 6.2.2.1).

The main memory of IBM's 370-158 model released in 1973 consisted of 1,024-bit NMOSFET SRAM cells.

2.3. DRAM memories

Central memories developed using MOSFET technology are volatile memories, as they do not retain information when the power is cut.

2.3.1. Intel 1103 cell

The first DRAM cell, AMS 6001, was developed and marketed by Advanced Memory Systems, a company founded in 1968 by defectors from IBM and Fairchild. The cell featured four to six transistors. The 1 kbit cell was sold to numerous companies: Honeywell, Raytheon, Wang Computer, etc.

In 1969, Honeywell asked Intel, a company founded in 1968 by Fairchild defectors, to develop a DRAM cell based on a three-transistor cell. This Intel 1102 chip had numerous problems, and Intel secretly set about developing a 1103 cell (1 k DRAM PMOS 3T Si gate) (Figure 2.9), where the storage component is a Q_1 transistor. The Q_2 and Q_3 transistors enable access to the write and read lines. The read operation is nondestructive. Capacitor C is a parasitic capacitor.

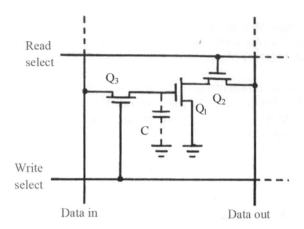

Figure 2.9. *DRAM cell. Intel 1103 1 k DRAM PMOS 3T Si gate (Lojek 2007, p. 362). For a color version of this figure, see www.iste.co.uk/vignes/silicon2.zip*

Memories based on this cell were intended to replace magnetic memories ferrite cores in computer memories from 1971 onwards.

2.3.2. *The one transistor DRAM cell*

According to Chih-Tang Sah, "the impact of the invention of the one-transistor DRAM cell is comparable to that of the invention of the transistor itself", "a worldwide consensus" (Sah 1988).

In 1963, IBM decided to develop and manufacture the NMOSFET transistor, which was to be one of the components of a DRAM memory cell (one transistor DRAM cell) designed by Robert Dennard in 1968 (Dennard 1968). The basic cell contains one NMOSFET transistor, which acts as a switch, and a MOS capacitor on which the information (1 bit) is stored in the form of an electrical charge (Figure 2.10). Figure 2.11 shows a cross-section of an elementary cell produced by Hitachi in 1983. The structure of such a circuit will evolve considerably over time.

An amplifier (thresholding amplifier) is associated with the base cell, as variations in the capacitor's charge generate very small voltage variations, which must be amplified to be detected. In addition, the signal must be restored after each reading, and as the capacitor leaks, it must be recharged in the absence of a reading.

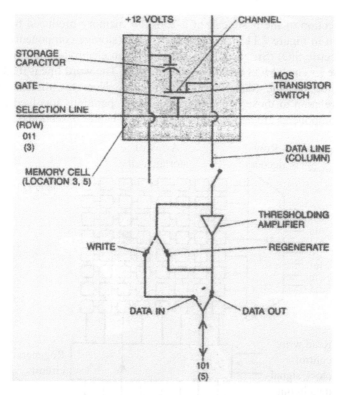

Figure 2.10. *One transistor DRAM cell (Hodges 1977, p. 57)*

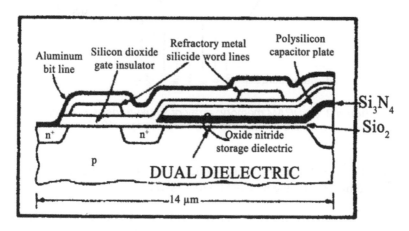

Figure 2.11. *Cross-section of a DRAM cell with an NMOS transistor (Sah 1988, p. 1288)*

A cross-section of the basic cell of a 256 kbit memory produced by Hitachi in 1983 is shown in Figure 2.11. The capacitor (right) (storage component) consists of a mixed dielectric SiO_2 /Si_3 N_4 polycrystalline silicon electrode, the other electrode being the base P. The NMOS transistor is on the left. The word line is made of metal silicide and the bit line (data line) of aluminum. Note the dimensions of the capacitor in relation to those of the transistor. The operations involved in building the various components are described in Volume 1, Chapter 6, section 6.2.2.

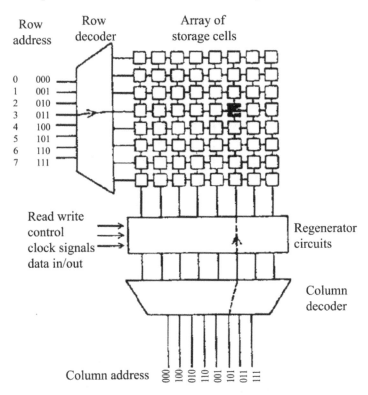

Figure 2.12. *DRAM memory: array of single-transistor DRAM cells accessible via two address lines (Hodges 1977, p. 57)*

In 1973, the DRAM memory: array of one-transistor DRAM cells 1024 and 2048 was developed by IBM for the new IBM 370/158-168.

In 1972, Intel replaced its 1103 memory with the 2104 memory (4 kbit DRAM NMOS 1T Si gate), which is identical in design to the IBM memory. The first mass production of memory chips widely used in computers was the memory manufactured by Intel (Rideout 1979).

In 1973, the Mostek company, founded by Texas Instruments defectors, marketed a DRAM memory made up of cells identical to those of IBM, but accessible via only two address lines (Figure 2.12), thus reducing the number of address lines required by a factor of 2 (compare Figures 2.9 and 2.12). The Mostek MK4116 DRAM 16 kbit memory, launched in 1976, took 75% of the world market (Hodges 1977).

In the early 1980s, the manufacture of these memories was monopolized by the Japanese. Annual production was expected to reach 100 billion cells by 1985.

2.4. Storage memories

The information is retained even when the power is cut.

2.4.1. *ROM memories*

Read-only memory (ROM) contains data and programs. Memory is a matrix of diodes or transistors. Each elementary cell placed at the intersection of a row and a column consists of a diode or a transistor. A "one" or "zero" is represented by the presence or absence of a diode or transistor.

PROM programmable memories (Figure 2.13) consist of ROM cells with a diode or transistor and a fuse that can be destroyed by passing a high current through the corresponding transistor, enabling the user to irreversibly fix the memory contents. When column i and line j are selected, if the fuse is intact, the transistor passes and imposes a logic 0 on the output, otherwise it is blocked and the output is reduced to a 1. The fuse is made of polycrystalline silicon or a metal alloy.

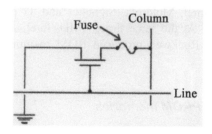

Figure 2.13. *Structure of a programmable cell (Douillard 2006, p. 57)*

The structure of a 3 × 3 bit NMOS programmable ROM is shown in Figure 2.14(a), whose contents are represented by the truth table (Figure 2.14(b)).

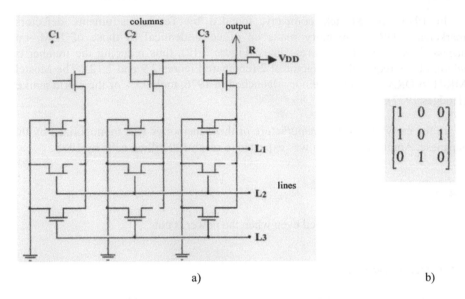

Figure 2.14. *(a) Structure of an NMOS ROM memory of 3*3bits (Douillard 2006, p. 56); (b) corresponding truth table*

In 1965, ROM memories appeared on the market.

Sylvania produces a 256-bit ROM memory (16 × 16 matrix of elementary cells) of TTL bipolar transistors for Honeywell, programmed bit by bit.

General Microelectronics produces a ROM memory 1,024-bit 32 × 32 array of MOSFET transistor cells.

In 1970, Fairchild, Intel, Motorola, Signetics and TI produced 1,024-bit TTL ROM bipolar memories. At the same time, AMD, Intel, AMI, Electronic Arrays, General Instrument and Rockwell produced ROM memories based on MOSFET transistors.

2.4.2. EPROM and EEPROM memories

2.4.2.1. History

It has been shown (Volume 1, Chapter 6) that the first MOSFET was designed and built by Ian Ross in 1955. The field effect was achieved by a gate consisting of a metal electrode and a dielectric on the germanium P base of an N-P-N bipolar structure (Volume 1, Chapter 6, Figure 6.7). The dielectric insulating the grid from the base was a ferroelectric material. In fact, Ian Ross's device was the cell of a nonvolatile memory.

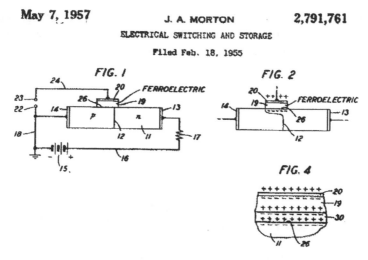

May 7, 1957   J. A. MORTON   2,791,761

ELECTRICAL SWITCHING AND STORAGE

Filed Feb. 18, 1955

Figure 2.15. *Component consisting of a PN diode and a capacitor forming a memory cell (Morton 1957)*

A memory cell based on this structure by Morton of Bell Labs is shown in Figure 2.15 (Morton 1957). It consists of a PN diode, a capacitor (electrodes 20 and 26) and the gate dielectric (19). As the dielectric is a ferroelectric material, the application of an electric field by the gate electrode (20) induces a remanent electric polarization of the dielectric (30) (orientation of the dipoles parallel to the electric field), creating a double layer that remains after the electric field has been removed (figure 4 in Figure 2.15). Polarization of the dielectric (30) induces the formation of an inversion channel on the underlying surface of a PN diode (figure 2 in Figure 2.15). This dielectric polarization inducing the formation of an inversion channel (26) could be "read" by the passage of a modulated current through the diode in the absence of a voltage applied to the gate. Conversely, in the absence of prior polarization of the ferroelectric layer, and therefore of a voltage applied to the gate, there is no inversion channel (and the component behaves like a simple diode).

2.4.2.2. The floating-gate MOSFET transistor

In 1967, Dawon Kahng and Simon Sze of Bell Labs designed and produced a new component for a nonvolatile storage cell: the floating gate MOSFET memory (Kahng and Sze 1967) (Figure 2.16), with the following structure:

Al (control gate) (19)/ZrO_2 layer (18)/Zr (floating gate) (17)/thin insulating SiO_2 layer (5 nm) (16)/n-Si (substrate).

This structure is based on the idea of using an electrical charge stored in an insulated metal gate (floating gate) of a MOSFET structure on a permanent basis (10 years or more) to store one "bit" of information.

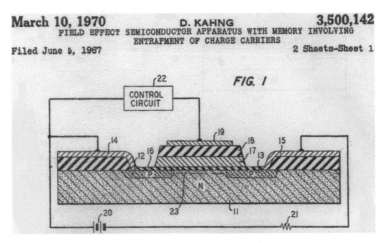

March 10, 1970   D. KAHNG   **3,500,142**
FIELD EFFECT SEMICONDUCTOR APPARATUS WITH MEMORY INVOLVING
ENTRAPMENT OF CHARGE CARRIERS
Filed June 5, 1967   2 Sheets–Sheet 1

Figure 2.16. *The MOSFET transistor with floating gate and control gate (Kahng 1970)*

2.4.2.3. *Operating principle of the MOSFET transistor with floating gate and control gate*

Dawon Kahng's invention is a silicon-based N (1 ohm.cm) MOSFET with silicon P source and drain islands and two gates: a floating metal (zirconium) gate (17) (storage gate, completely insulated, separated from the base by a 5 nm silica insulating layer) (16) and a control gate (19) (metal), which plays the same role as the gate of a MOSFET transistor. The separation between the floating gate and the control gate is a 100 nm layer of ZrO_2 (18). By applying a positive voltage to the control gate (19), of the order of 20–50 V, under the effect of the electric field created, the electrons of the N-base (11) go by tunnel effect (Volume 1, Chapter 2, section 2.1.2.1) into the floating (insulated) electrode (17). To achieve this, the thickness of the insulating layer separating the N-base (11) from the insulated grid (17) must be very thin of the order of 5 nm. The write pulse of around 20–50 V and its duration of 0.5 µs enable 5×10^{12} electrons/cm^2 to be stored on the floating gate. As the floating gate (17) is electrically insulated, electrons injected into the gate are trapped. When the floating gate is charged with electrons, it acts as a screen for the electric field between the control gate (19) and the N-base (11), and therefore increases the threshold voltage of the transistor (the threshold voltage V_{th} is the voltage above which a conductive inversion channel p (23) is formed in the N base between source p and drain p (Volume 1, Chapter 6, Figures 6.1 and 6.2). This means that a higher threshold voltage $V_{th2} > Vt_{h1}$ must be applied to the control gate (19) for a current to flow between source and drain when the floating gate is loaded than when the floating gate is unloaded. When an intermediate voltage $V_{th1} < V_{in} < V_{th2}$ is applied to the control gate (19), a current flows between source and drain; this means that the floating gate is not loaded, so a logic "one" is stored in the floating gate (17). For the same intermediate voltage applied to the control gate (19), if no current flows between source and drain, the floating gate is charged. As no current flows, a logic "zero" is stored in the floating gate. The presence of a logic "zero" or "one" is therefore detected by applying a voltage intermediate to the threshold voltage $V_{th1} < V_{in} < V_{th2}$ to the control gate (19).

Derivatives of the floating-gate transistor have been developed.

2.4.2.4. EPROM memory

The FAMOS transistor (floating gate avalanche injection mos) is the component of an EPROM/UV-PROM memory, invented by Froman-Bentchkowsky (1971) (Figure 2.17).

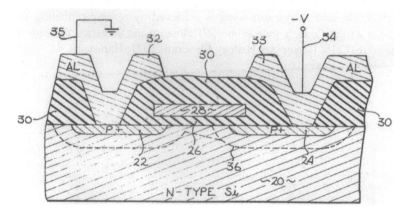

Figure 2.17. *FAMOS transistor: single floating gate transistor (Froman-Bentchkowsky 1972)*

The FAMOS is an N-substrate, P-channel PMOSFET device with a single gate, the floating gate, made of polycrystalline silicon (28), fully insulated and embedded in silica (SiO_2 insulating layers 30/1 μm and 26/50 to 100 nm). The transistor behaves like an open switch. Storage in the floating gate is achieved by tunneling. As the gate is insulated, negative charges remain stored when the drain-source voltage is reduced to zero. The negative charges stored in the gate create an electric field, and the transistor enters the inversion regime. An inversion channel p is formed in the substrate. The transistor then behaves like a closed switch.

Erasing is achieved by subjecting the component to UV radiation of the order of 10 W/cm^2 for a few minutes. The photoelectric effect extracts the stored electrons from the grid.

The industrial development of this EPROM memory was undertaken by Intel, who in 1971 marketed a 2 kbit memory (erasable programmable read-only memory), eventually called UV-PROM (ultraviolet light erasable programmable read-only memory).

2.4.2.5. EEPROM, flash memory

The EEPROM (electrically erasable nonvolatile semiconductor memory) cell (Figure 2.18), designed by Harary and based on the Kahng cell (Figure 2.16), is a p-channel MOS transistor with two gates: a fully insulated polycrystalline silicon floating gate (56) embedded in silica and a metallic control gate (62). Storage in the floating gate is achieved by avalanche injection of hot electrons, when a positive voltage is applied to the control electrode (62), with a current flowing between source and drain. Erasing is achieved by applying a negative or zero voltage to the control electrode, and charge removing is achieved by reverse tunneling from the narrow zone of the floating gate (zone 59) close to the substrate through the thin dielectric zone (CHM (Computer History Museum) 1971; Harary 1978).

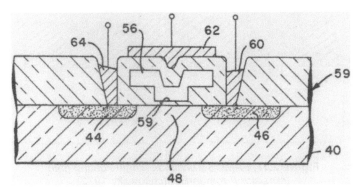

Figure 2.18. EEPROM memory floating gate transistor (Harary 1978)

In 1978, Intel marketed EEPROM (electrically erasable programmable read-only memory) (CHM (Computer History Museum) 1971). Fujio Masuoka, at Toshiba in 1980, designed the flash memory (Masuoka and Iizuka 1985) (figure 6 in Figure 2.19).

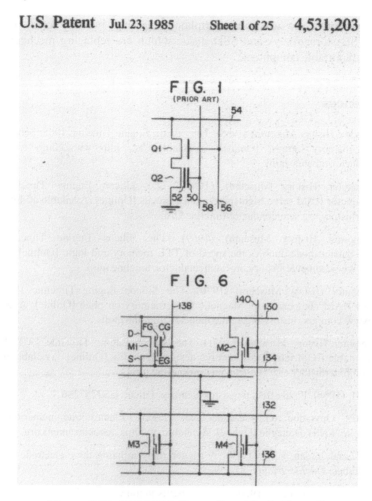

Figure 2.19. *EEPROM (figure 1) and flash (figure 6) memory cell architectures (Masuoka and Iizuka 1985)*

In an EPROM memory, the cells are made up of a single transistor (1T cell structure). All cells are erased at the same time. In an EEPROM memory, cells are made up of two transistors (one for cell selection and one for storage (2T cell structure)). Each cell can therefore be erased when selected (figure 1 in Figure 2.19). A flash memory cell has the same structure as an EEPROM, but the cell is a single-transistor structure. Memory capacity is therefore greater (figure 6 in Figure 2.19).

Flash memories are used in smartphones, tablets, memory cards in digital cameras, USB storage keys and SSD disks, which are replacing mechanical hard drives in ultraportable computers.

2.5. References

CHM (Computer History Museum) (1965). The Silicon Engine Timeline 1965: Semiconductor read-only-memory arrays [Online]. Available at: http://www.computerhistory.org/semiconductor/timeline.html.

CHM (Computer History Museum) (1966). The Silicon Engine Timeline 1966: Semiconductor RAM serve high-speed storage needs [Online]. Available at: http://www.computerhistory.org/semiconductor/timeline.html.

CHM (Computer History Museum) (1969). The Silicon Engine Timeline 1969: Schottky-barrier diode doubles the speed of TTL memory and logic [Online]. Available at: http://www.computerhistory.org/semiconductor/timeline.html.

CHM (Computer History Museum) (1970). The Silicon Engine Timeline 1970: MOS Dynamic RAM competes with Magnetic Core memory on price [Online]. Available at: http://www.computerhistory.org/semiconductor/timeline.html.

CHM (Computer History Museum) (1971). The Silicon Engine Timeline 1971: Reusable Programmable ROM introduces iterative design flexibility [Online]. Available at: http://www.computerhistory.org/ semiconductor/timeline.html.

Dennard, R.H. (1968). Field effect transistor memory. Patent, US3387286.

Douillard, C., Ouvradou, G., Jézéquel, M. (2006). Électronique numérique/logique séquentielle. ENST Bretagne [Online]. Available at: https://coursexamens.org.

Eccles, W.H. and Jordan, F.W. (1919). A trigger relay utilizing three-electrode thermionic vacuum tubes. *The Electrician*, 83, 298.

Eccles, W.H. and Jordan, F.W. (1920). Improvements in ionic relays. Patent, GB148582.

Faber, A.S. and Schlig, E.S. (1967). Non-destructive memory array. Patent, US3354440.

Frohman-Bentchkowsky, D. (1971). Floating gate transistor and method for charging and discharging. *Applied Physics Letter*, 18(4), 332–334.

Frohman-Bentchkowsky, D. (1972). Floating gate transistor and method for charging and discharging. Patent, US3660819.

Harary, E. (1978). Electrically erasable non-volatile semiconductor memory. Patent, US4115914.

Harris, J.D. (1958). Direct-coupled transistor logic circuitry. *IEEE Transactions on Electronic Computers*, 1(3), 2–6.

Hodges, D.A. (1977). Microelectronic memories. *Scientific American*, 237(3), 54–63.

Kahng, D. (1970). Field effect semiconductor apparatus with memory involving entrapment of charge carriers. Patent, US3500142.

Kahng, D. and Sze, S.M. (1967). A floating gate and its application to memory devices. *Bell System Technical Journal*, 46(4), 1288–1295.

Lojek, B. (2007). *History of Semiconductor Engineering*. Springer, Berlin.

Masuoka, F. and Iizuka, H. (1985). Semiconductor memory device and method for manufacturing the same. Patent, US 4531203.

Moore, G. (1995). Interview with Gordon Moore (March 3). Silicon Genesis website [Online]. Available at: http://www-sul.stanford.edu/depts/hasrg/histsci/silicongenesis/moore-ntb.html.

Morton, J.A. (1957). Electrical switching and storage. Patent, US2791761.

Norman, H. (1971). Solid state switching and memory apparatus. Patent, US3562721.

Pugh, E.W. (1981). Solid state memory development in IBM. *IBM Journal of Research and Development*, 25(5), 585–602.

Rideout, V.L. (1979). One device cells for dynamic random access memories. A tutorial. *IEEE Transactions Electron Devices*, 26(6), 839–852.

Sah, C.T. (1988). Evolution of the MOS transistor. *Proceedings of the IEEE*, 76(10), 1283–1326.

Schlig, E.S. (1981). Multiple access store. Patent, US4280197.

Schmidt, J.D. (1965). Integrated MOS transistor random access memory. *Solid State Design*, 1, 21–25.

Seitz, F. and Einspruch, N.G. (1998). *The Tangled History of Silicon*. University of Illinois Press, Champaign.

Shannon, C. (1938). A symbolic analysis of relay and switching circuits. *Transactions of the American Institute of Electrical Engineers*, 57(12), 713–723.

Shannon, C. (1948). A mathematical theory of communication. *Bell System Technical Journal*, 27(3), 379–423.

Sze, S.M. (2002). *Semiconductor Devices*, 2nd edition. John Wiley & Sons, New York.

Liquid Crystal Displays

The liquid crystal display (LCD) (Figure 3.6) is made up of an array of cells. Each cell consists of a pixel, filled with liquid crystals and a type of resistor called the thin film transistor (TFT).

It is the TFT, a direct field effect transistor, made from a thin film of amorphous silicon, developed in 1979, which enabled the development of flat LCDs.

It was not until 1989 that the first commercial 10.4" TFT/LCD screen was launched.

3.1. The TFT – A history

3.1.1. *The Lilienfeld direct field-effect transistor*

The TFT, with a Cu_2S film, was "invented" by Julius Lilienfeld, who filed a design patent on March 28, 1928 for a "Device for controlling electric current". This transistor (Figure 3.1) consists of a gate electrode, an aluminum plate (10) covered with a dielectric layer of alumina (produced in situ) (11) (about 10^{-4} mm thick) and a semiconducting Cu_2S film (12), as well as source (14) and drain (15) metal electrodes. The semiconductive properties of Cu_2S had been known since 1851: "An extremely intensive electric field may be established in a minute thickness of this layer [...] The conductivity of such coating or layer of minute thickness will depend upon the electrostatic field applied across the insulating layer". It is not known whether such a component was actually manufactured and tested. If it had, no effect would have been observed, due to the "trap" problem (Lilienfeld 1933a).

On the same day, March 28, 1928, Lilienfeld (1933b) filed a design patent for an "electrical condenser device", consisting of three layers of Al/Al$_2$O$_3$/Cu$_2$O or Cu$_2$S.

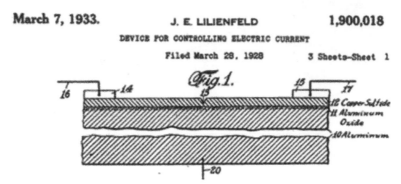

Figure 3.1. *Lilienfeld patent for a direct field-effect transistor (Lilienfeld 1933a)*

3.1.2. *Welker's TFT*

A TFT with a copper oxide Cu$_2$O film was manufactured by Heinrich Welker in Munich and tested in March 1945. It consisted of an insulating support plate, a thin layer of copper oxide (whose semiconducting properties had been known since 1927), 10 μm thick, an insulating layer of dielectric 200–500 μm thick, a metal plate at positive potential, creating a strong electric field in the semiconductor layer, and two electrodes in contact with the semiconductor layer, between which the current flows through the semiconductor layer. "Only small effects" were observed. The result was negative, as the dielectric layer was too thick (Welker 1945; Riordan and Hoddesson 1997, pp. 112–114).

3.1.3. *Bardeen, Brattain and Gibney's research on the direct field effect*

The research carried out by Bell Labs as early as 1946 had already demonstrated this direct field effect but the effect obtained on their components was very weak. This was due to traps at the semiconductor/dielectric interface for the majority carriers (electrons) (Volume 1, Chapter 3, section 3.1.2).

Numerous experiments on the direct field effect were carried out by Pearson as early as 1945 (reported in 1948). The component consisted of a slice of fused quartz 75–100 mm thick, with a surface area of 1 × 2 cm, coated on one side with a film of germanium N film on one side and a gold deposit on the other (Volume 1, Chapter 3,

Figure 3.3). A high voltage applied to the gold deposit (the gate) caused an increase in the current flowing through the semiconductor film (measured between two gold electrodes (source and drain)) deposited on the semiconductor film. Measurements were made on films of germanium P, silicon N and Cu_2O films. These experiments showed only a very weak effect, contrary to theory. With the Cu_2O film, an increase in relative conductivity of 0.11 was obtained for an electric field of 400,000 V/cm (V = 3,000 V). "Although the modulation of 0.11 is not great, the useful output power is substantial. It is in principle operative as an amplifier. This is a moral victory", wrote Pearson in his laboratory book (reported by Riordan and Hodesson, 1997, pp. 143 and 315).

It was the experiments by Bardeen, Brattain and Gibney at Bell Labs, in November–December 1947, that demonstrated a direct field effect on an experimental device consisting of a thin layer of germanium P, of very low thickness, formed on a block of germanium N. The conductivity of this thin layer was modulated by an electric field of negative polarity, that is, a direct field effect (Volume 1, Chapter 3, section 3.1.2.2 and Figure 3.7; see also Volume 1, Chapter 6, section 6.1.5).

3.1.4. *Weimer's direct field-effect transistor*

In 1961, the first TFT based on field effect control of majority carriers was patented by Paul Weimer of the RCA company (Weimer 1966).

As indicated in Volume 1, Chapter 6, following the oral presentation by Bell Labs researchers of their MOSFET transistor, RCA, not wanting to miss out on a potential opportunity pursued developments on the MOSFET transistor. At the same time, it entrusted a recently recruited researcher with the task of developing a transistor that could compete with the MOSFET.

In the late 1950s, RCA was one of the companies heavily involved in research into semiconductor materials, particularly CdTe and CdS for solar cells (Chapter 4, section 4.1), deposited in thin films.

According to Paul Weimer, the idea that complex circuits could be built from thin-film transistors, which are much cheaper than "solid" silicon, was in the air ("there was a thought"). Weimer's CdS transistor (Figure 3.2) consisted of thin layers deposited on a glass substrate: a first layer of two gold or copper electrodes (12 and 14) spaced 20 mm apart, a second layer of CdS N (16), a third layer of an insulating material (SiO_2, CaF_2) (18) of very low thickness (<2 mm) and a fourth metal layer (Au, Cu) constituting the control electrode through which the voltage creating the electric field is applied (20). With this device and the very thin films constituting the respective insulator and semiconductor layers, the interface traps (see Volume 1, Chapter 3, section 3.1.2) are saturated by majority carriers (here, electrons) for a positive potential applied to the grid > 6 V. Beyond this, an increase in gate potential produces an accumulation of charges near the semiconductor/dielectric interface, and thus an increase in conductivity and current flow in the semiconductor layer between source and drain (Weimer 1962).

However, the transistor was never developed further due to the difficulties of manufacturing it on large surfaces.

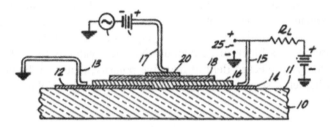

Figure 3.2. *Weimer (1966) direct field-effect transistor*

3.2. The amorphous silicon TFT

It was not until 1979 that the amorphous silicon TFT, which was to ensure the development of liquid crystal displays, was manufactured. This followed on from the solar energy developments undertaken by the US Department of Energy in 1974.

The first thin-film amorphous silicon solar cells were manufactured by RCA in 1976 (Chapter 4, section 4.7).

The first silicon film TFT, consisting of a 1-μm-thick film of hydrogenated amorphous silicon (a-Si:H) and a 0.7-μm-thick layer of dielectric Si_3N_4, was manufactured in 1979 using the plasma-enhanced chemical vapor deposition (PECVD) process (see Figure 3.4) by Le Comber and Spear (1979) who were involved in the development of amorphous silicon solar cells since 1975.

Figure 3.3 shows cross-sections of MOSFET and TFT transistors, showing the a-Si thin film (in red).

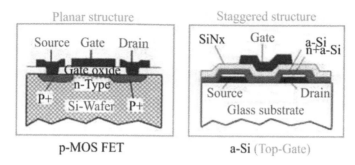

Figure 3.3. *Typical MOSFET and a-Si TFT structures (AVDEALS; Tixier-Mita et al. 2016). For a color version of this figure, see www.iste.co.uk/vignes/silicon2.zip*

The thin-film a-Si is hydrogenated amorphous silicon a-Si:H. Hydrogen ions block the dangling bonds (Volume 1, Chapter 1, section 3.1.2.1), allowing electrons to flow normally through the thin film (see Chapter 4, section 4.7). Two doped regions n+ a-Si provide an ohmic contact with the source and drain (Volume 1, Chapter 2, section 2.1.2).

The two basic structures of TFTs are shown in Figure 3.4. These transistors are to be implemented in a TFT-LCD cell exposed to light radiation (Figure 3.8), as a-Si:H amorphous silicon is a photovoltaic material, the transistor needs to be shielded from any light rays to avoid generating a disruptive current.

This is why the bottom gate TFT configuration transistor is used in TFT-LCD cells. The gate's metal electrode (in blue) and the SiN_x insulator film block the light.

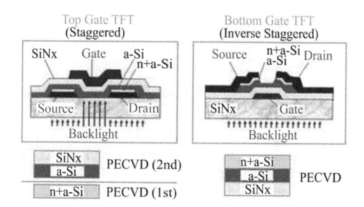

Figure 3.4. *The two basic structures of TFT (AVDEALS). For a color version of this figure, see www.iste.co.uk/vignes/silicon2.zip*

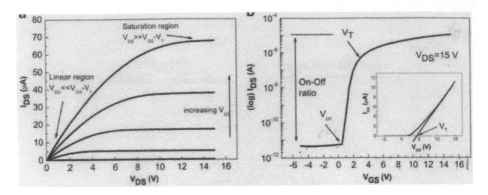

Figure 3.5. *Characteristic curves of
an n-type TFT (Correia et al. 2016)*

The characteristic curves (current flowing through the transistor as a function of source-drain voltage V_{DS} and source-gate voltage V_{GS}) of a TFT are shown in Figure 3.5. Note the high operating voltages of this transistor. In the ON position, the drain-source voltage V_{DS} is of the order of 15 V, the voltage applied to the gate V_{GS} is 15–20 V, and in the OFF position it is negative, of the order of – 5 V. In both linear and saturated regions, the expressions for the current I_{DC} as a function of the source-drain voltage V_{DS} and source-gate voltage V_{GS} are those of a MOSFET transistor (Volume 1, Chapter 6, section 6.1.2.2, formulas [6.4] and [6.5], and Figure 6.3).

The basic characteristics of the a-Si:H TFT are as follows:

– electron mobility: $\mu_{eff} < 1$ cm^2/V·s (compare with electron mobility in electron silicon: 1,400 cm^2/V·s);

– saturation current (off current): $< 10^{-12}$ A;

– threshold voltage: $V_T < 3$ V.

These characteristics are sufficient to switch a liquid crystal display that operates at a frequency of 120 Hz (Brody 1984).

Amorphous silicon makes it "easy" to deposit large areas of silicon at relatively low temperatures (with the PE CVD process). What is more, it can be deposited on any transparent substrate (glass or plastic) without surface limitation, hence, its use in liquid crystal flat-panel displays (section 3.3, Figure 3.8).

These transistors are also used in X-ray detectors as sensors, marketed at the end of the 1990s, for diagnosing fractures, lesions, etc. The transistor's support surface is covered with an X-ray photoconductor such as CdTe. A dose of X-rays produces an electrical charge that is stored in a capacitor integrated into the TFT array.

3.3. Liquid crystal displays

3.3.1. *History*

The first LCD was designed and produced by a team led by George Heilmeier at RCA between 1964 and 1968 (RCA had built its fortune on television) (Heilmeier 1970, 1976).

In the same years, in the laboratories of the Westinghouse company, James Fergason designed and built a liquid crystal display screen measuring 10 × 12 inches.

At the end of 1969, Fergason (1973), who had left Westinghouse in 1966, founded ILIXCO and discovered the twisted nematic (TN) cell.

The first LCD screen, 6 × 6 inches, 20 lines per inch, based on TN-LCD technology and driven by a CdSe film TFT, was designed, built and demonstrated by Brody et al. (1973) at Westinghouse in 1973.

While the amorphous silicon a-Si:II film TFT was developed in 1979, it was not until 1989 that the first commercial 10.4" TFT/LCD screen was launched.

3.3.2. *Structure and operation of a liquid crystal display*

The flat-panel liquid crystal display is a passive device. It does not emit light. It transfers polarized light. The screen operates with LED backlighting. Only its transparency varies, so it must be backlit.

The structure and operation of the screen are briefly presented for information.

The liquid crystal display (Figure 3.6) is made up of an array of cells. Each cell consists of a pixel and a TFT (9). The pixel is located between two external plates, the "polarizers" (2 and 3) and two electrodes: a front electrode (10) made of a transparent, conductive ITO (indium tin oxide) film and an array of rectangular rear electrodes (11) (defining the size of the cell) made up of ITO. The two ITO films were deposited on two glass plates (1) spaced a few micrometer apart (5 μm).

The cavity between the two electrodes is filled with liquid crystals, controlled by the electrodes (10 and 11), acting on the orientation of the crystals (Figure 3.7).

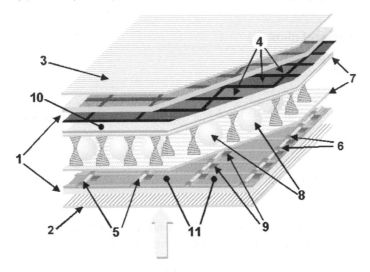

Figure 3.6. *Liquid crystal display (AMLCD) (Wikipedia n.d.). For a color version of this figure, see www.iste.co.uk/vignes/silicon2.zip*

Helical nematic crystal displays (TN-LCD) discovered in 1971 by Schadt and Helfrich, uses the principle of light polarization (Schadt and Helfrich 1972).

The operation of a cell is shown in Figure 3.7. The cell consists of two polarizer plates (outer plates) and two electrodes (inner plates), and it is filled with liquid crystals (blue).

Figure 3.7(a), in the absence of an electric field (OFF state), the orientation of liquid crystals (in blue) describes a quarter-helix from one electrode to the other. This arrangement has the property of rotating the polarization direction of an incident light beam (green). The polarization of light by the first polarizer (lower outer plate (2)) follows the variation in orientation of the liquid crystals and exits the cell in the same orientation as the second polarizer (upper outer plate (3)). The cell thus transmits the light polarized by the input polarizer.

Figure 3.7(b) (ON state), by applying an electric field between the current-carrying (ITO) electrodes (inner plates (10) and (11)), the liquid crystal molecules orient themselves parallel to this field, perpendicular to the glass substrates. The polarization of incident light remains unchanged through the liquid crystal. The light is absorbed by the second polarizer crossed at 90°.

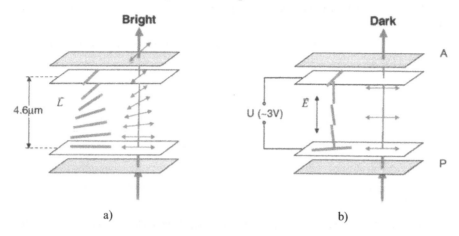

Figure 3.7. *Structure and operation of a liquid crystal cell: (a) no voltage between inner plates; (b) under electric field (Fujitsu Microelectronics 2006). For a color version of this figure, see www.iste.co.uk/vignes/silicon2.zip*

The OFF (bright) and ON (dark) states are controlled by the TFT (9).

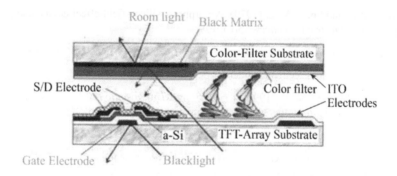

Figure 3.8. *Structure of a Samsung LCD cell (AVDEALS n.d.). For a color version of this figure, see www.iste.co.uk/vignes/silicon2.zip*

Figure 3.8 shows a schematic cross-section of a cell (not to scale) featuring the TFT on the left, the part of the cell through which the light passes and, on the far right, a capacitor to maintain voltage. The yellow plates are the ITO electrodes.

3.4. References

AVDEALS (n.d.). TFT-LCD TFT device design [Online]. Available at: http://www.avdeals. com.

Brody, T.P. (1984). The thin film transistor – A late flowering bloom. *IEEE Transactions on Electron Devices*, 31(11), 1614–1628.

Brody, T.P., Asars, J.A., Dixon, G.D. (1973). A 6 x 6 inch 20 lines-per-inch liquid-crystal display panel. *IEEE Transactions on Electron Devices*, 20(11), 995–1001.

Correia, A.P.P., Barquinha, P.M.C., Da Palma Goes, J.C. (2016). Thin film transistors. In *A Second-Order ΣΔ ADC Using Sputtered IGZO TFTs*, Correia, A.P.P., Barquinha, P.M.C., Da Palma Goes, J.C. (eds). Springer International Publishing, Cham [Online]. Available at: http://www.springer.com/978-3-319-27190-3.

Dacey, G.C. and Ross, I.M. (1953). Unipolar "field-effect" transistor. *Proceedings IRE*, 41, 970–979.

Fergason, J. (1973). Display devices utilizing liquid crystal light modulation. Patent, US3731986.

Fujitsu Microelectronics (2006). Fundamentals of liquid crystal displays – How they work and what they do. White paper, Fujitsu Microelectronics America, Inc.

Heilmeier, G. (1970). Liquid crystal display devices. *Scientific American*, 222(4), 100.

Heilmeier, G. (1976). Liquid crystal displays: An experiment in interdisciplinary research that worked. *IEEE Transactions on Electron Devices*, 23, 780.

Le Comber, P.C. and Spear, W.E. (1979). Amorphous-silicon field-effect device and possible applications. *Electronics Letters*, 15(6), 179.

Lilienfeld, J.E. (1933a). Device for controlling electric current. Patent, US1900018.

Lilienfeld, J.E. (1933b). Electrical condenser device. Patent, US1906691.

Riordan, M. and Hoddeson, L. (1997). *Crystal Fire: The Invention of the Transistor and the Birth of the Information Age*. Norton & Company, New York.

Schadt, M. and Helfrich, W. (1972). Lichtsteuerzelle. Patent, CH532261.

Shockley, W. and Pearson, G.L. (1948). Modulation of conductance of thin films of semi-conductor by surface charges. *Physical Review*, 74, 232.

Tixier-Mita, A., Ihida, S., Ségard, B.D., Cathcart, G.A., Takahashi, T., Fujita H., Toshiyosh, H. (2016). Review on thin-film transistor technology, its applications. *Japanese Journal of Applied Physics*, 55, 04EA08.

Weimer, P.K. (1962). Solid state device with gate electrode on thin insulative film. *Proceedings of the IRE*, 1462–1469.

Weimer, P.K. (1966). Solid state device with gate electrode on thin insulative film. Patent, US3258663.

Welker, H. (1945). Halbleiteranordnung. German Patent, 980084.

Solar Cells

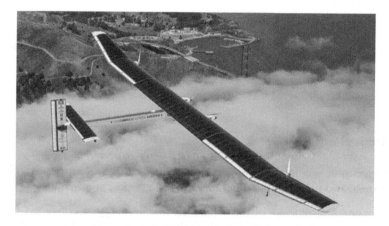

Figure 4.1. *Solar impulse: 17,248 monocrystalline silicon cells, 135-μm thick, solar conversion efficiency 23% (source: https://stonybrookstories.files.wordpress.com/2015/ 03/solar-impulse-2-b1.jpg). For a color version of this figure, see www.iste.co.uk/vignes/ silicon2.zip*

Silicon is the "material" of solar cells: 93% of terrestrial modules with a conversion efficiency >20%.

A solar cell consists essentially of a PN diode.

The photovoltaic effect of a silicon PN diode was discovered in 1940, along with the rectifier effect of such a junction.

The solar cell development program began in 1952 at Bell Labs, to meet the needs of ATT (Bell Labs' parent company), which was looking for sources of power for its telephone network in remote areas. On April 26, 1954, Bell Labs announced the manufacture of silicon solar cells with an efficiency of 6%.

The space program was the driving force behind the development of these cells (the *Vanguard* satellite was launched in 1958).

The energy crisis of 1974–1975 relaunched the cell development program.

Increasing cell efficiency and reducing costs are the driving forces behind the manufacturing technology development programs.

This chapter presents:

– some figures on solar power generation;

– the discovery of the photovoltaic effect presented by a silicon PN junction;

– the basics of crystalline and amorphous silicon solar cell operation;

– photovoltaic silicon manufacturing technologies.

4.1. Introduction

Optoelectronic components (interaction between light energy and electrical energy) are of two types: emitters and receivers (detectors) of radiation:

– Light-emitting diodes (LEDs) where electrical energy is converted into light energy. This is the photoemissive or electroluminescent effect. LEDs emit light by electron–hole recombination, yielding photons (Volume 1, Chapter 1, section 1.2.5). Only direct-transition semiconductors exhibit this effect. The semiconductors used are GaAlAs, GaInAsP on an InP substrate, ZnSe and InGaN. The width of a semiconductor's energy bandgap determines the wavelength of the light emitted. Silicon and germanium, as indirect-transition semiconductors, do not exhibit this photoemissive effect, as electron–hole recombinations generate heat.

– Solar cells and photodiodes where radiation energy is converted into electrical energy. This is the photovoltaic effect. Although solar cells capture all or a large fraction of the solar spectrum, photodiodes are designed for high sensitivity (quantum efficiency) in a narrow band of the electromagnetic spectrum.

In fiber optic telecommunications, at the entrance of the fiber, LED converts the electric signal into light energy and at the end of the fiber, photodiodes convert the light energy into an electrical signal. At the entrance of the fiber, GaAlAs over GaAs substrate LEDs are used for 850 nm wavelength and GaInAsP over InP substrate for wavelength (1,300 nm). At the end of the fiber, silicon photodiodes are used for 850 nm wavelength transmissions. For wavelengths in the infrared range (1,300–1,550 nm), which correspond to the dispersion and attenuation minima of silica fibers, InGaAs/Si or InGaAs/InP photodiodes are used because of their high sensitivity in the 800–1,700 nm range.

4.2. Silicon: the material of solar cells

Solar cells convert solar energy directly into electrical energy (photovoltaic effect).

In 2015, silicon was the material of 93% of the terrestrial solar cell modules produced, in three forms: thick monocrystalline (mono-Si) (24%), thick polycrystalline (multi-Si) (69%) and thin film (amorphous silicon a-Si) (a few %). The variation in annual module production since the 2000s is shown in Figure 4.2. The market share of amorphous silicon cells has been declining since 2011 (Figure 4.3). Conversion efficiencies of thick-film cells are higher than or equal to those of thin-film cells (Figure 4.4).

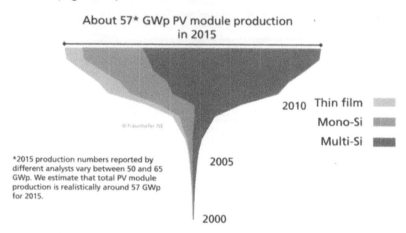

Figure 4.2. *Annual global module production since the 2000s in cumulative GWp (gigawatts of peak power) (Fraunhofer ISE 2016). For a color version of this figure, see www.iste.co.uk/vignes/silicon2.zip*

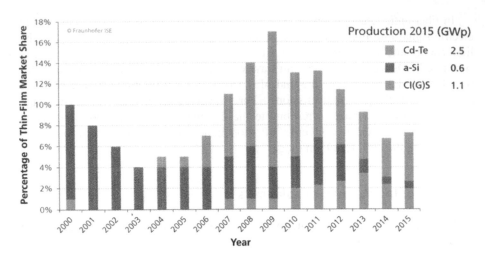

Figure 4.3. *Market share of thin-film cells as a percentage of total production (Fraunhofer ISE 2016). For a color version of this figure, see www.iste.co.uk/vignes/silicon2.zip*

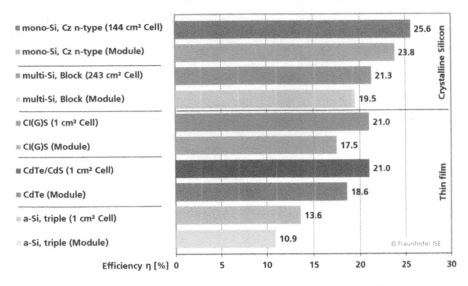

Figure 4.4. *Conversion efficiency of cells and modules of different technologies (Fraunhofer ISE 2016). For a color version of this figure, see www.iste.co.uk/vignes/silicon2.zip*

It should be noted that germanium does not appear as an earth solar cell material in these graphs. Solar conversion efficiency (Figure 4.13) is only 9%, due to the small energy bandgap E_G, as we will see later. What is more, the cost of the base material precludes its use for solar cells.

The same is true for GaAs, although its conversion efficiency is 26% due to its very high cost. In space, on the other hand, it is the basic material for triple-junction cells, consisting of GaInP/GaAs/Ge connected cells on germanium substrates, which achieve conversion efficiencies of 43.5% (Fraunhofer ISE 2016) and open-circuit voltages of 2.26 V at costs of $45,000/m^2$.

As for thin-film solar cells CdTe/CdS, CIS, CIGS (CuInGaSe) and amorphous Si (a-Si), their slow development can be explained by the following reasons:

– CdTe has the advantage of great stability over time and moderate cost, but two problems: tellurium availability and cadmium toxicity;

– copper/indium/selenium (CIS), copper/indium/gallium/selenium (CIGS) and copper/indium/gallium/diselenide/disulfide (CIGSS) offer the highest yields among thin films, but at a higher cost;

– amorphous silicon (a-Si) (<1%) has a low conversion efficiency.

4.3. History[1]

In 1839, Edmond Becquerel presented a *mémoire sur les effets électriques produits sous l'influence des rayons solaires* to the Académie des Sciences, in which he showed that an electric current was established in "a cell containing two different liquids" when these were exposed to light (Becquerel 1839). In 1817, Jacob Berzelius discovered selenium and showed that it was a non-conductor, unlike tellurium. In 1851, Hittorf proved that selenium is conductive at ordinary temperature in one of its allotropic varieties, and that its electrical resistance decreases with temperature up to the melting point, demonstrating that selenium is a semiconductor (Hittorf 1851). The photoconductivity (where electrical conductivity increases with the intensity of light irradiation) of gray selenium (a semiconductor) was discovered by Willoughby Smith in 1873 (Smith 1873) and by Werner von Siemens in 1875, who designed the first photometer made of selenium film (Siemens 1875). In 1877, Adams and Day discovered the photovoltaic effect (which they called the photoelectric effect[2]) presented by a selenium–platinum junction. They found that when the junction was connected to a battery, current flowed in one

---

1 A complete history of solar cells is given by F.M. Smits of Bell Labs (Smits 1976).

2 The photoelectric effect (discovered by Hertz in 1887) consists of the emission of electrons from a metal surface subjected to electromagnetic radiation. Einstein was awarded the Nobel Prize in Physics in 1921 for his work in theoretical physics, and in particular for his discovery of the laws of the photoelectric effect.

direction. By disconnecting the junction from the battery and subjecting it to light, they found that a current flowed in the opposite direction: "The first demonstration that electricity could be produced from light without heat or moving parts" (Adams and Day 1877). In 1883, Fritts (1883) manufactured the first functional selenium cell, the selenium layer being covered by a thin layer of gold, with an energy yield of 1–2%. This photovoltaic effect was later used to create light meters (photoelectric cells) in the 1930s (Kodes 1971). The inventor of the telephone, Alexander Graham Bell, invented the "photophone" in 1880. He used a selenium cell to detect the light beam modulated by a sound wave.

Selenium is used in copper/indium/selenium (CIS) and copper/indium/gallium/ selenium (CIGS) solar cells (Figure 4.4).

The photovoltaic effect presented by a silicon PN junction drawn from a silicon ingot was discovered by Russell Ohl of Bell Labs in 1940 (Ohl filed two patents in 1941) (Volume 1, Chapter 4, section 4.1.1) (Ohl 1946, 1948). It should be remembered here that when the remarkable photovoltaic property of this junction was brought to the attention of the director of Bell Labs, Mervin Kelly decided that this discovery was of great value to the electronics industry, and that absolute secrecy should be maintained until further study revealed its full power.

The solar cell development program began in 1952 at Bell Labs. David Chapin, to meet the needs of ATT (Bell Labs' parent company), was looking for sources of power for the telephone network in remote areas. At the same time, Calvin Fuller was looking for processes to produce PN junctions. Gerald Pearson who was studying the manufacture of power diodes, components with large PN contact areas, using the diffusion doping process (Volume 1, Chapter 5, section 5.2.4), found that these diodes were "sensitive" to light. Pearson and Chapin subjected these components to sunlight and observed a photovoltaic effect, with an efficiency of 4%. These first PN junctions were produced by diffusing lithium in silicon P plates. However, due to the high diffusivity of lithium in silicon, these components were unstable. At the end of 1953, Fuller used boron diffusion (p dopant) in silicon N wafers to produce large-area PN junctions with a strong photovoltaic effect and a conversion efficiency of up to 6%. In March 1954, Calvin Fuller filed his first patent on the diffusion doping process for boron doping of silicon N (Fuller 1962). On April 26, 1954, Bell Labs announced the manufacture of silicon solar cells using the diffusion doping process (described in Volume 1, Chapter 5, section 5.2.4) (Chapin et al. 1954, 1977).

ATT's first rural telephone relay system, experimentally powered by solar cells, was installed in Georgia in October 1955. A panel of 432 silicon solar cells supplying 9 W in full sunlight was fixed to the top of a pylon (Figure 4.5). For

6 months, the system operated continuously and satisfactorily. For economic reasons, however, the experiment was not pursued.

Figure 4.5. *Solar batteries and Western Electric type 1858 transistor. The first solar panel operated by Bell Telephone Laboratories in 1954 (Warren 2016)*

The space program was the driving force behind the development of solar cells. The first use of solar cells was on the *Vanguard 1* satellite, launched on March 17, 1958, to power a radio transmitter, which operated for 8 years (Figure 4.6). The space program stimulated (financed) a great deal of research and a real cell production industry.

Figure 4.6. *The Vanguard satellite (NASA, National Space Science Data Center). For a color version of this figure, see www.iste.co.uk/vignes/silicon2.zip*

As far as research is concerned, as early as 1955, a fundamental relationship between the energy bandgap width E_G (Volume 1, Chapter 1, section 1.2.2) of the semiconductor and the efficiency of a solar cell was established by Prince (1955), leading Loferski (1956) to show that the optimum energy bandgap for a solar conversion efficiency of 20% should be 1.6 eV. As the energy bandgap of silicon is 1.2 eV, this result prompted a great deal of research into different semiconductors (GaAs, InP, CdTe and CdS) in the late 1950s, particularly by RCA (Radio Corporation of America) (Chapter 3, section 3.2). In 1956, researchers at Wright-Patterson Air Force Base (Reynolds et al. 1954) in the United States published results on a Cu_2S/CdS solar cell with an efficiency of 6%. A year later, a GaAs solar cell with 6% efficiency was announced by researchers at RCA (Jenny et al. 1956). GaAs, in monocrystalline thin-film form, enables high efficiency at the cost of great complexity (cascaded photovoltaic cells), with conversion efficiency reaching 47.1% in 2019. But the cost of manufacturing limits its application in space. A thin-film CdTe solar cell with conversion efficiencies of 6% was manufactured by Cusano (1963).

But as none of these materials performed significantly better than silicon in terms of solar yield, silicon became the basic material. An additional circumstance contributed to silicon's supremacy. Satellites (the main use for solar cells at the time), having to operate at high altitude, were exposed to nuclear radiation (the Van Allen belt contains a significant flux of high-energy protons). It was observed that the lifetime of NP solar cells exposed to nuclear radiation was three times longer than that of PN cells. The first *Telstar* telecommunications satellite equipped with such NP cells was launched on July 10, 1962.

It was the energy crisis of 1974–1975 that sparked renewed interest in solar cells. At the time, it was estimated that a reduction in cost by a factor of 100 was needed to make the production of electrical energy using this technology economically viable. Two programs to develop manufacturing processes for "polycrystalline silicon" (see Volume 1, Chapter 4, section 4.2.2.2) were funded by the Jet Propulsion Laboratory (JPL/LSA program) and the US Department of Energy (ERDA).

4.4. Solar radiation and its absorption

4.4.1. *Solar radiation*

The energy of radiation emitted by the sun as a function of wavelength, the solar spectrum, is shown in Figure 4.7. When a semiconductor is subjected to light radiation of energy $E_v = hv$ greater than its band gap, $E_v > E_G$ (see Volume 1, Chapter 1, section 1.2.2, Table 1.1 and Figure 1.7), each absorbed photon creates an electron–hole pair. The generation of electron–hole pairs is described in Volume 1, Chapter 1, section 1.2.5.

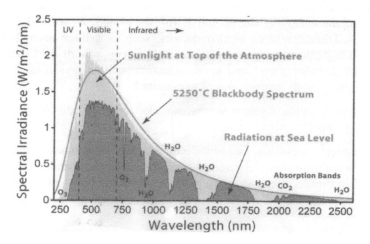

Figure 4.7. *The solar radiation spectrum: at the top of the atmosphere (yellow), at sea level (red) and the blackbody spectrum at 5,250°C (solar surface temperature) (credit: Nick84 (CC BY-SA 3.0), via Wikimedia Commons). For a color version of this figure, see www.iste.co.uk/vignes/silicon2.zip*

The energy bandgap E_G of a semiconductor is the minimum energy that the radiation (photon) must possess in order to be absorbed. The smaller the bandgap, the more the semiconductor absorbs long-wave (low energy) light, which makes up a large proportion of sunlight (Figure 4.7). Germanium therefore appears to be the material that can absorb the most light. Silicon, with a bandgap $E_G = 1.2$ eV, absorbs wavelengths up to 1,100 nm in the infrared, while GaAs, with a bandgap 1.43 eV, only absorbs wavelengths below 800 nm. Amorphous silicon, $E_G = 1.6$ eV, absorbs in the 400–700 nm band.

4.4.2. *Absorption of light energy*

Photons entering a P semiconductor block travel variable distances x before being absorbed. Absorption of a photon generates an electron–hole pair. For an incident photon flux $\Phi_0(E)$ (number of photons of energy E striking the surface unit and penetrating the material per second), the fraction of photons absorbed at a distance x is proportional to the intensity of the residual flux at that distance:

$$d\Phi(x)/dx = -\alpha\ \Phi(x)$$

where $\Phi(x)$ is the residual flux at x. Radiation attenuates exponentially as it propagates, which defines an absorption coefficient α ($1/\alpha$ is the penetration depth below which all energy is absorbed). This coefficient depends on the type of material and the energy of the photon E (of wavelength λ (μm) $= 1.24/E$ (eV)) (Figure 4.8). Direct-transition semiconductors, such as gallium arsenide (GaAs), copper indium diselenide and amorphous a-Si:H silicon, absorb light strongly. On the other hand, indirect-transition semiconductors such as crystalline silicon are weakly absorbing, with the exception of germanium:

– in direct-transition semiconductors, an electron–hole pair is generated by absorbing a photon;

– in indirect transition semiconductors[3], two phenomena must occur simultaneously for a valence electron to enter the conduction band: absorption of a photon and absorption or emission of a phonon, depending on whether the energy of the photon hν is less than or greater than the width of the energy band gap.

---

3 This includes direct and indirect band gap materials. Given the complexity of the phenomena involved in photon absorption, a simple explanation is almost impossible. The interested reader may wish to consult basic books on semiconductor physics. In this section, the expression "direct or indirect transition material" does not have the same meaning as in Volume 1, Chapter 1, section 1.2.3, where electron–hole generation/re-combination is referred to as indirect, since it takes place in two stages.

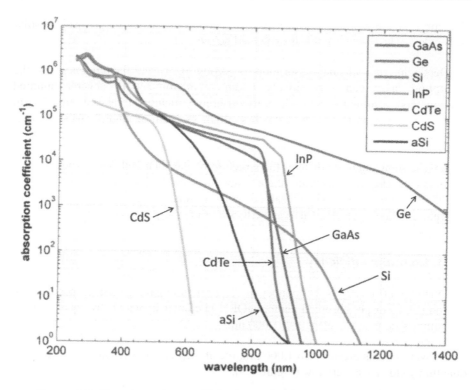

Figure 4.8. *The absorption coefficient as a function of radiation wavelength for various semiconductors at 300 K (Honsberg and Bowden n.d.). For a color version of this figure, see www.iste.co.uk/vignes/silicon2.zip*

For direct-transition semiconductors, the absorption coefficient varies according to a stair-step law (Figure 4.8):

$$\alpha = 0 \text{ for } E_{photon} < E_{Gap}; \text{ and } \alpha = \text{constant for } E_{photon} > E_{Gap}$$

α is of the order of 10^5/cm for GaAs, InP, CdTe semiconductors, whereas the absorption coefficient for an indirect transition semiconductor (Si, Ge) decreases continuously with the wavelength (Si curve for silicon, Ge curve for germanium) (Figure 4.8).

Germanium has two ideal characteristics for solar energy conversion: a low bandgap of 0.72 eV, which means it absorbs radiation down to the infrared range, and an absorption coefficient of solar energy greater than all other semiconductors

(Ge curve in Figure 4.8). However, the conversion efficiency of germanium solar cells is low. The reason for this is explained below.

Depending on whether absorption is strong or weak, the thickness of the absorption layer can vary greatly (from a micrometer to several hundred micrometers), resulting in cells with very different structures. It takes around 50 μm of crystalline silicon to absorb as much sunlight as 1 μm of amorphous silicon, or 0.1 μm of copper indium diselenide.

So, the semiconductor dictates the structure of the solar cell. Two types of cells are being developed: cells with "thick" substrates for indirect transition materials, essentially mono- and polycrystalline silicon, and "thin-film" cells made of amorphous silicon (a-Si:H), CdTe and CIGS (CuInGaSe$_2$).

4.4.3. Quantum efficiency

Quantum efficiency is the number of electron–hole pairs generated per incident photon (EQE) or per absorbed photon (IQE). The latter is practically equal to 1 for crystalline silicon at wavelengths <1,000 nm.

At wavelengths greater than 1,000 nm, the absorption coefficient is very low, the quantum yield is also very low (Figure 4.9).

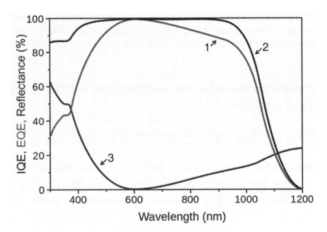

Figure 4.9. *Quantum efficiency of crystalline silicon as a function of radiation wavelength per incident photon (curve 1) and per absorbed photon (curve 2). Curve 3 is the reflectivity (Wikipedia: "quantum efficiency"). For a color version of this figure, see www.iste.co.uk/vignes/silicon2.zip*

4.5. Crystalline silicon solar cells

4.5.1. *Manufacturing technology*

A solar cell (Figure 4.10(a)) consists of a mono- or polycrystalline substrate, with a thickness of between 135 and 300 μm of a silicon P, lightly doped with p boron ($N_A = 10^{15}$ atoms/cm^3); on the rear side of the substrate: a conductive layer, a metal that serves as an ohmic contact; on the front side: a thin layer of N$^+$ semiconductor, heavily doped with phosphorus ($N_D = 10^{20}$ atoms/cm^3), with a thickness <1 μm (0.35 μm), exposed to radiation (photons pass through without being absorbed); an anti-reflective layer and an electrical contact grid.

The N$^+$ layer is produced by the diffusion process (described in Volume 1, Chapter 5, section 5.2.4) by decomposition of the gaseous compound POCl$_3$ on the substrate surface and diffusion of phosphorus into the substrate surface layer constituting the thin N$^+$ layer with formation of a phosphosilicate layer (PSG) on the N$^+$ layer.

This operation extracts the iron (a harmful impurity) present in the surface layer of the P base, the "gettering effect", whose atoms concentrate in the phosphosilicate layer (described in section 4.6.2).

4.5.2. *Physical basis of operation*

A solar cell is essentially a large-area diode made up of two regions of N$^+$ and P semiconductors, whose juxtaposition produces a depletion layer, in which the electric field is located (Figure 4.10(b)) (see Volume 1, Chapter 4, section 4.1.2).

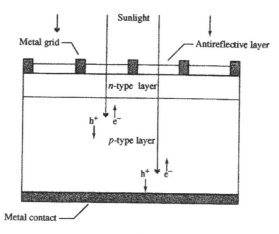

a)

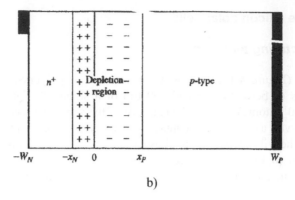

b)

Figure 4.10. *Solar cell: N⁺P diode. (a) Creation of an electron–hole pair e⁻ and h⁺,*
respectively, in the P layer by a photon; (b) depletion layer created by the
migration of P-layer free carriers into the N⁺ layer (Gray 2003, pp. 62 and 84)

The electron–hole pair generated by photon absorption (Volume 1, Chapter 1, section 1.2.5), essentially in the P region, in the vicinity of the N⁺ layer, is separated by the electric field created in the depletion layer (Figure 4.10(b)). Electrons generated in the P layer (p-type) with energy E_p cross the depletion layer and enter the N⁺ layer. The number of electrons generated per unit time I_{gen} is proportional or equal to the number of photons absorbed per unit time, that is, the light intensity I_L. The diode behaves like a generator.

If the two sides of the cell are connected by a conductor (with resistance R_L), the charges generated by the radiation flow into the external circuit. The voltage V across the cell increases with the current produced:

$$V = R_L.I_{gen} \hspace{3cm} [4.1]$$

With the usual sign conventions, this current flows in the diode in the N → P (direction) (electrons flow in the P → N direction), that is, it flows in the opposite direction to the current flowing in a PN diode (see Volume 1, Chapter 4, Figure 4.3(b)).

But the electron generated at a certain depth in the P region must travel the corresponding distance in this P layer to reach the N layer, and it may undergo the phenomenon of "electron–hole recombination" (Volume 1, Chapter 1, section 1.2.5) as a result of the presence of "traps" constituted by metallic impurities and lattice defects (grain boundaries, dislocations) in this P layer and in the depletion region.

This distance is characterized by the diffusion length: a function of the lifetime of the electrons generated (see Volume 1, sections 4.1.2, 4.1.4 and 4.3.2, formulas [4.7a] and [4.10] and Figure 4.4). A fraction of the electrons generated in the P layer by solar absorption is therefore trapped by electron–hole recombination in this same P layer and in the depletion zone. The current I produced by the solar cell is therefore the result of electron–hole pair generation and recombination of electron–hole pairs in the P region.

The lifetime of electrons generated by radiation in the P region is a function of the concentration of holes introduced by the p-type dopants in this region (boron) (N_a /cm^3). The higher the dopant density, the greater the probability of electron–hole recombination. For this reason, the P region of solar cells is lightly doped with boron.

Standard solar cells operate in a range of $n_{(P)}$ (excess carrier density) minority carrier concentrations (free electrons "generated" in the P substrate) from 1×10^{13} to 5×10^{14}/cm^3, proportional to light intensity. Solar cells with high conversion efficiencies operate in a minority carrier concentration range of 1×10^{15} to 1×10^{16}/cm^3.

4.5.3. Characteristic curve

The characteristic curve of a photodiode and a solar cell (Figure 4.11(a)) is that of a diode (see Volume 1, Chapter 4, Figure 4.2), translated by an amount corresponding to the current generated by irradiation I_{SC}.

The operating range of a *photodiode* is the lower left quadrant. A photodiode is a diode subjected to solar radiation and reverse polarity voltage, and the current measured is the current generated by the irradiation I_{SC}.

The operating range of a *solar cell* or *photovoltaic cell* is in the lower right quadrant. The operating point is that of a circuit consisting of a photovoltaic cell and a load R_L fed by it. It is given by the intersection of the diode "characteristic" with the load line with slope $1/R_L$ (formula [4.1]).

The characteristic is generally presented inverted around the potential axis (Figures 4.11b and 4.12). The generated current I_{gen} is composed of two terms: the current generated by irradiation I_{SC} (photocurrent) and the current of a PN under the bias V fixed by the external resistor (Volume 1, Chapter 4, section 4.5.2, formula [4.13]):

$$I_{gen} = I_{SC} - I_{s.}\{exp(eV/kT) - 1\} \qquad [4.2]$$

where $V = R_L$. I_{gen} (formula [4.1]) and I_S is the saturation current of the diode (Volume 1, Chapter 4, formulas [4.13] and [4.15]):

$$I_S = en_i^2 (D_n/N_A L_n) \qquad [4.3]$$

For a low external resistance R_L and therefore a low voltage V, the diode current is low, and the resulting current is close to the current created by irradiation (photocurrent):

$$I_{gen} = I_{SC} \qquad [4.4]$$

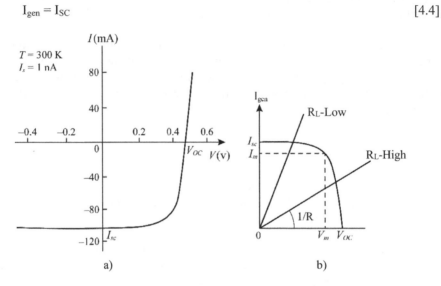

a)   b)

Figure 4.11. *(a) Current–voltage characteristic curve (for A = 4 cm²); (b) "inverted" current representation and operating points (Sze 2002, p. 321)*

If the resistance R_L is increased, the voltage V increases, producing an increase in the diode current, and the resulting current decreases. At high resistance and high voltage, the resulting current becomes zero. The corresponding voltage V is V_{OC} (open circuit) (Figure 4.11(b)). Figure 4.12 shows typical values. A typical characteristic curve with typical solar cell parameters.

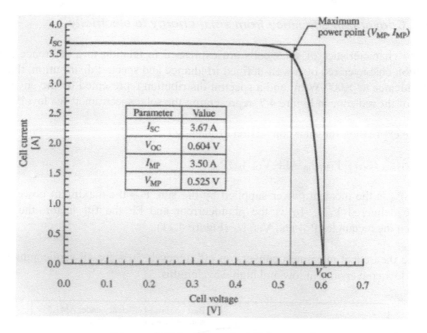

Figure 4.12A. *Characteristic curve of a silicon solar cell (A = 100 cm²)*

Parameter	Value
A	100 cm²
W_N	0.35 μm
N_D	1×10^{20} cm⁻³
D_p	1.5 cm²/V-s
$S_{F,eff}$	3×10^4 cm/s
τ_p	1 μs
L_p	12 μm
W_P	300 μm
N_A	1×10^{15} cm⁻³
D_n	35 cm²/V-s
S_{BSF}	100 cm/s
τ_n	350 μs
L_n	1100 μm

Figure 4.12B. *Corresponding parameters (Gray 2003, p. 93)*

4.5.4. *Conversion efficiency from solar energy to electricity*

The characteristics of solar cells are expressed in relation to a reference solar radiation characterized by a well-defined irradiance and spectral distribution, that is, an irradiance of 1,000 W/m² and a spectral distribution represented by the envelope curve of the red zone in Figure 4.7, representing the solar spectrum at sea level[4].

The expression for solar conversion efficiency is:

$$\eta_{max} (E_G) = P_{MP}/P_{in} = FF.V_{OC} \, I_{SC}/P_{in} \qquad\qquad [4.5]$$

where P_{in} is the incident power supplied by the sun, P_{MP} the maximum power that can be delivered $V_m I_m$, I_{SC} is the photocurrent and FF the fill factor, the ratio between the rectangles P_{MP} and $V_{OC} \, I_{SC}$ (Figure 4.11).

The theoretical maximum conversion efficiency is 30% for all semiconductors, due to low conversions at low and high wavelengths.

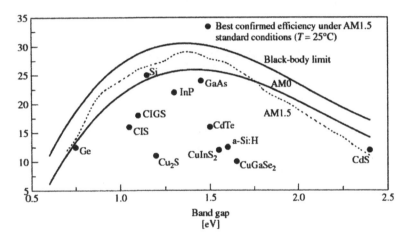

Figure 4.13. *Solar energy conversion efficiency as a function of the semiconductor's energy band gap (Goetzberger and Hebling 2000; Bailey and Raffaelle 2001)*

---

4 Just above the atmosphere, solar radiation intensity is 1,358 W/m². The water vapor, ozone and carbon dioxide in the atmosphere absorb some of the sun's energy. By the time light reaches sea level, its power is down to 1,000 W/m² (irradiance). At the Earth's surface, the wavelengths of the solar spectrum range from 0.4 to 2.5 μm, with a maximum at around 0.50–0.55 μm.

The conversion efficiencies of different cells are shown in Figure 4.13 (values achieved in 2001, to be compared with those in Figure 4.4, of the same order of magnitude).

Crystalline silicon solar cells offer the best performance. An energy bandgap of 1.1 eV is considered optimal. The cells absorb light at all wavelengths between 0.35 and 1 μm. Energy band gaps between 1.1 and 1.6 μm give materials their optimum conversion efficiencies[5]. By contrast, amorphous silicon solar cells (a-Si:H) (with an energy bandgap of the same order of magnitude) have lower efficiencies, due to their disordered structure.

4.5.4.1. *The case of germanium*

Although germanium has two ideal characteristics for this conversion, a low bandgap (0.72 eV), enabling the absorption of solar radiation down to the infrared range, and the highest absorption coefficient (Figure 4.8 Ge curve) of all semiconductors, solar energy conversion efficiency is low, at around 15%. This is because the open-circuit voltage V_{OC} of such a diode is very low (the highest measured value is 0.229 V), whereas for a silicon cell, the voltage V_{OC} reaches 0.6 V and even 0.7 V (Figure 4.12) for a generated current $I_L = I_{SC}$ of the same order of magnitude. The voltage V_{OC} is a function of the ratio I_{gen}/I_S and therefore, for the same generated current, a function of the saturation current I_S of a PN diode:

$$V_{OC} = kT/e \log (I_{SC}/I_S)$$
[4.6]

As the saturation current I_S (formula [4.3]) is proportional to the square of the intrinsic carrier density n_i^2, germanium has a high intrinsic charge carrier density ($2.8 \times 10^{13}/cm^3$) (Volume 1, Chapter 1, Table 1.1). The "saturation current" I_S of a germanium diode is particularly high: 1,000 times higher than for silicon, hence the low voltage V_{OC}.

4.6. Action of metallic impurities on the solar conversion efficiency

4.6.1. *Lifetimes and diffusion lengths of "electrons": influence of doping. The gettering effect*

The parameter controlling solar cell performance, that is, solar conversion efficiency, is the lifetime (Volume 1, Chapter 4, formula [4.7a]) or the diffusion length (Volume 1, Chapter 4, formula [4.10]), in the P-base of the cell, of the

---

5 Recent conversion efficiencies for CIGS and CdTe solar cells are close to optimum values.

electrons "generated" by the absorption of a photon (Figure 4.10). The diffusion length must be equal to or greater than the thickness of the base, so that free electrons generated at the end of the base do not recombine with holes by the mechanism described in Volume 1, Chapter 1, section 1.2.5.1, Figure 1.8. The optimum thickness for a solar cell is between 60 and 100 µm. The thickness of the monocrystalline solar cells on the *Solar Impulse* aircraft is 135 µm.

The influence of impurities (oxygen, carbon, metallic elements), lattice defects (grain boundaries, dislocations, etc.) on the lifetime and diffusion length of minority carriers is presented in Volume 1, Chapter 4, section 4.1.4.

For solar cells, the content of metallic impurities, mainly transition metals, is iron (as it comes from the raw material: metallurgical silicon). It is the main factor controlling this diffusion length and therefore the degradation of conversion efficiency.

4.6.2. *Monocrystalline cells*

Two sets of results characterize the influence of metal impurity content, one on diffusion length or lifetime of electrons "generated" in a P-base, and the other on the solar conversion efficiency cells whose monocrystalline wafers have been obtained using the CZ process, that is, on bases that are extremely pure and virtually free of lattice defects.

The variation in lifetime and diffusion length of an electron in a P base as a function of iron content (depending on whether the iron is in the state of an interstitially dissolved atom or in the state of a FeB compound) in a single crystal obtained by the CZ process is shown in Figure 4.14 (already given in Volume 1, Chapter 4, Figure 4.8).

For monocrystalline solar cells with thickness >100 µm, the concentration of iron in the P-base of the cell must be $<10^{12}$ atoms/cm^3 for Fe$_i$ and $<10^{13}$ atoms/cm^3 for FeB in order to obtain a diffusion length >100 µm.

The second set of results concerns the influence of metal content (Figure 4.15) on the solar conversion efficiency of monocrystalline solar cells. It decreases from the following concentrations: for iron, $>10^{14}$ atoms/cm^3, that is, 2 or 4 ppbm; for titanium, $>10^{11}$ atoms/cm^3.

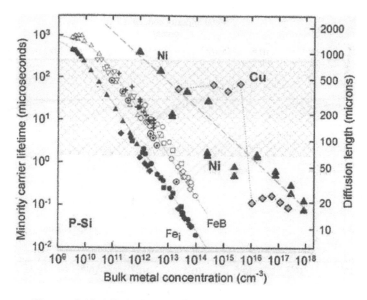

Figure 4.14. *Lifetimes and diffusion lengths of Fe, Ni, Cu impurities as a function of their concentrations in wafers taken from ingots obtained by the CZ process (Istratof et al. 2006)*

The results for the diffusion length of electrons in the P base (Fe < 10^{12} a/cm³) (Figure 4.14) and those relating to cell conversion efficiency (Fe < 10^{14} a/cm³) (Figure 4.15) are discordant. In fact, the iron concentration values reported in Davis' study (Figure 4.15) are those measured on wafers making up the P base prior to cell manufacture. However, the manufacture of the PN^+ cell includes a phosphorus diffusion doping operation using the CVD-diffusion process (Volume 1, Chapter 5, section 5.2.4.1) to produce the N^+ thin film (at 850°C, 50 min) (Dastghelb-Shirazi et al. 2013). This operation simultaneously extracts the iron present in the surface layer of the P base by the "gettering effect": iron atoms are concentrated in the phosphosilicate layer (PSG) that forms on the N^+ layer. After this operation, the iron concentration in the P-base decreased sharply, as confirmed by numerous measurements. The iron content in the P base (after completion of the cell) ensuring an adequate diffusion length (>150 µm) is therefore of the order of Fe < 10^{11} atoms/cm³. But the iron content in the P wafer prior to solar cell production must be <10^{14} atoms/cm³. The "gettering effect", due to the oxidation treatment of silicon, leading to purification of the silicon, had been observed and exploited in the development of point-contact diodes for radar reception (Volume 1, Chapter 2, section 2.3.1.3).

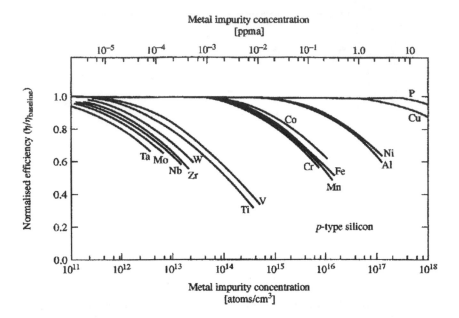

Figure 4.15. *Solar cell conversion efficiency as a function of impurity concentration in a 4 Ω-cm P-base solar cell from CZ ingots (Davis et al. 1980)*

4.6.3. *Polycrystalline cells*

Polycrystalline cells, made from wafers drawn from ingots obtained by unidirectional solidification (processes described below: Figure 4.17), present lattice defects (grain boundaries, dislocations) (more or less numerous depending on the drawing process) on which impurity atoms, particularly iron atoms, which are supersaturated in the lattice, come to segregate. These Fe-rich zones (clusters, etc.) are particularly active sites for electron–hole recombination in the cell's P base.

The production of the thin N+ layer by phosphorus doping, which simultaneously produces the extraction of iron, the gettering effect, has a greater or lesser efficiency in extracting the iron atoms segregated on the lattice defects.

Coletti et al.'s (2008) study on unidirectionally solidified ingot slices (section 4.8.2) having undergone the iron extraction treatment shows that:

– for slabs from the reference ingot with iron content $<10^{11}$ a/cm^3, the electron lifetime of the electrons in the P base of the cells, before and after the making of the N$^+$ layer, is homogeneous along the length of the ingot and of the order of 100 µs (in agreement with the results shown in Figure 4.14), and conversion efficiencies are of the order of 20% along the length of the ingot (Figure 4.4);

– for ingot slabs with an iron content $\approx 10^{14}$ a/cm^3, without iron extraction treatment, the lifetime value is homogeneous along the length of the ingot, but very low, of the order of 1 μs, confirming the iron effect. After iron extraction, (gettering effect) only the wafers drawn from the center of the ingot (between 40 and 70% of the length) show relatively high lifetimes, of the order of 50 μs, and conversion efficiencies of just under 20%.

As a result, for wafers from unidirectionally solidified ingots, and for iron contents $\approx 10^{14}$ a/cm^3, only a small fraction of the ingot yields cells with relatively good yields.

4.7. Amorphous silicon solar cells

The bandgap of amorphous silicon is 1.6 eV. This is optimal, making amorphous silicon highly sensitive in the visible light range.

But the atomic structure of amorphous silicon a-Si is disordered, and many of its dangling bonds constitute "traps", favoring electron–hole recombination (see Volume 1, Chapter 3, section 3.1.2).

Saturating the dangling bonds by hydrogen atoms of amorphous silicon a-Si:H reduces the density of "traps" (by a factor of around 1,000) and gives the material its semiconducting properties. Amorphous a-Si:H silicon has a very high light absorption coefficient (1,000 times that of crystalline silicon). A thickness of the order of a micrometer is sufficient to produce solar cells.

Amorphous silicon films a-Si:H are produced using the CVD deposition process using SiH$_4$ silane. The film contains around 10% hydrogen. The film can be deposited on relatively large surfaces on a variety of substrates (glass, etc.).

In 1969, Chittic et al. (1969) demonstrated the semiconducting properties of hydrogenated amorphous silicon, and in 1975 Spear and Le Comber showed that amorphous silicon films could be doped with phosphorus (n-type) or boron (p-type) by adding either the compound PH$_3$ or the compound B$_2$H$_6$ to the Si-H$_4$ gas (Spear and Lecomber 1975).

The first cells were manufactured by RCA in 1976, with efficiencies of around 5% (Carlson et al. 1977) (see Chapter 3, section 3.2). In the early 1980s, the Japanese (Sanyon, Kyocera, Fugy) launched production of low-efficiency photo-batteries (3%) for use in calculators.

In 1982, the Americans announced laboratory yields of 10%. Access to the energy market became likely. But low-cost manufacturing requirements, problems of stability over time and layer uniformity meant that the real yield of solar cells reached only 4% in 1984.

In 2015, the efficiency of an a-Si:H laboratory solar cell reached 13.6% (Figure 4.4), but the efficiency of industrial modules was only 8.1%. The market fraction occupied by such cells was 0.6% in 2015 (Figure 4.3).

NOTE.– The amorphous silicon TFT transistor was presented in Chapter 3.

4.8. Solar silicon manufacturing processes

The cost of solar cells is essentially that of the basic material, the silicon wafer.

Virtually all crystalline silicon wafers, that is, 93% of production in 2015, are made from electronic silicon (Si-EG) obtained by chemical purification processes (Siemens or Union Carbide processes) (presented in Volume 1, Chapter 4, section 4.2.2.2), and therefore they are of very high purity, especially in transition metals (purity 9 to 11N).

The process of manufacturing wafers for the electronics industry involves three stages: (1) purification of metallurgical silicon by chemical processes to produce silicon chloride (trichlorosilane) or ultrapure silicon hydride; (2) production of electronic (Si-EG) polycrystalline silicon, using the CVD process (decomposition of chlorides or hydrides on silicon rods in the Siemens process) (Volume 1, Chapter 4, Figure 4.12); (3) single-crystal pulling using the CZ process.

Hence, we search for either modifications to certain stages of the Siemens or Union Carbide processes, or purification processes using metallurgical operations (UMG: upgraded metallurgical silicon), or operations to produce crystalline structures using less expensive processes than the CZ process.

4.8.1. Production of polysilicon (Si-EG) in a fluidized-bed reactor

The cost of the final stage in the production of polysilicon (Si-EG) in rod form using the CVD process (silicon chloride $SiHCl_3 \rightarrow Si(s)$) in the Siemens reactor (Volume 1, Chapter 4, Figures 4.11 and 4.12) is very high due to the high energy consumption resulting from losses through the reactor walls.

Hence, the ongoing development by a number of companies of a process for producing granular polysilicon using the CVD process in a fluidized-bed reactor

FBR (Figure 4.16) (silicon deposition on particles serving as seeds, maintained in fluidization), while using as raw material silicon trichloride or silane purified by the Siemens or Union Carbide processes. This FBR-CVD process enables continuous production with an 80% reduction in energy consumption compared with the CVD process in the Siemens reactor. But technical difficulties (the reaction being endothermic, the need to heat the seed particles, silicon deposition on the walls and dust production) are delaying industrial implementation.

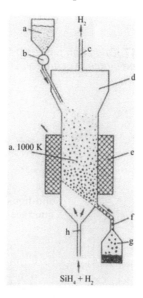

Figure 4.16. *FBR-CVD process: production of silicon powders in a fluidized-bed CVD reactor (20% SiH₄ + 80% H₂) at 1,000 K (REC Silicon 2011)*

4.8.2. *Production of columnar polycrystalline ingots*

The production of solar cells from wafers made from ingots of large columnar grains themselves made from electronic silicon (Si-EG) grains has been a major step forward in the quest for cost reduction; these ingots can be obtained using unidirectional solidification processes that are much less expensive than the CZ process. Indeed, when the ingot and therefore the polycrystalline wafers (multi-Si) are made from electronic silicon (Si-EG), whose content of metallic elements, particularly iron, is very low, lattice defects (dislocations, grain boundaries not decorated with iron) not being active recombination sites, the conversion efficiency of such cells reaches 21.3% (see Figure 4.4).

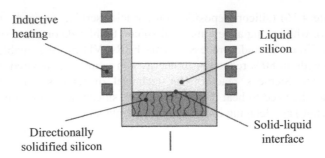

Figure 4.17. *Bridgman process. Columnar polycrystalline ingots (Koch et al. 2003, p. 215)*

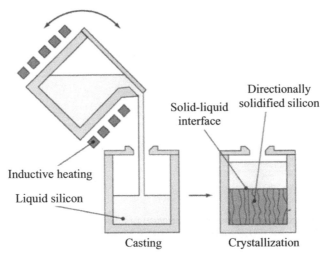

Figure 4.18. *Block casting process. Columnar polycrystalline ingots (Koch et al. 2003, p. 215)*

Columnar polycrystalline ingots are produced using either the Bridgman process (Figure 4.17), or the block casting process (Figure 4.18), which is much simpler than the CZ process, which also produces rectangular bricks. In the Bridgman process, unidirectional solidification is achieved by moving the pre-filled crucible relative to the hot zone. In the block casting process, solidification takes place as the ingot mold is filled. As with the CZ process, the starting material must be of very high purity, since following the high segregation of metallic impurities in the liquid phase as the ingot grows, the upper part of the ingot cannot be used.

4.8.3. *Purification of metallurgical silicon by metallurgical processes*

We present the process developed by the Elkem company in Norway using metallurgical silicon, whose major interest is the reduction in energy consumption, evaluated at 75% (Soiland et al. 2012).

It consists of three stages:

– pyrometallurgical stage: boron extraction by metal-slag reaction (B < 0.4 ppm);

– hydrometallurgical stage: leaching: extraction of phosphorus (30→3 ppmw, 90%) and metals (iron);

– directional solidification in the form of parallelepipeds to obtain rectangular wafers that can be used (without scrap) to make solar cells.

The chemical composition of ESS silicon (ESS brick) is as follows: B < 0.20 ppmw; P < 0.62 ppmw; C < 25 ppmw; Al < 0.01 ppmw; transition metals < 50 ppba; alkali metals < 200 ppba.

According to Elkem, the conversion efficiency of polycrystalline cells made from ESS silicon is around 16.5%.

This process presents two problems: the presence of boron and phosphorus and the iron content.

Metallurgical silicon purification processes are unable to reduce the content of two doping impurities of opposite polarity: boron and phosphorus. As a result, silicon is electrically compensated. But during silicon crystallization, due to the different segregation coefficients of the two main dopants (boron and phosphorus), the net doping concentration (difference in concentration between p-type acceptor dopant and n-type donor dopant) varies significantly along the ingot height, leading to a polarity inversion. Typically, at the bottom of the ingot, the silicon is p-type, then becomes n-type at the top of the ingot, leading to a significant reduction in material yield. In addition, there is a strong variation in n and p carrier density of more than an order of magnitude on the p-type part of the ingot, which has a negative impact on electrical performance.

With regard to iron, the effect of which on solar conversion efficiency and electron lifetime has already been presented (Figures 4.14 and 4.15), its content in the wafers varies along the length of the ingot, even after unidirectional solidification. Iron extraction during production of the N^+ layer of the PN^+ cell has a low, but above all highly variable efficiency, as we have seen in section 4.6.

These variations in composition along the length of the ingot produce a variation in resistivity and a change in conductivity type along the height of the ingot, which have a significant impact on material yield.

4.9. References

Adams, W.G. and Day, R.E. (1877). The action of light on selenium. *Philosophical Transactions of the Royal Society of London*, 25, 113–117.

Bailey, S. and Raffaelle, R. (2003). The history of space solar cells. In *Handbook of Photovoltaic Science and Engineering*, Luque, A. and Hegedus, S. (eds). John Wiley and Sons, Chichester.

Becquerel, E. (1839). Mémoires sur les effets électriques produits sous l'influence des rayons solaires. *C. R. Acad. Sci. Paris*, 9, 561.

Bell, A.G. and Tainter, S. (1880). Photophone. Patents, US23549, US235497, US235590.

Carlson, D.E., Wronski, C.R., Pankove, J.I., Zanzucchi, P.J., Staebler, D.L. (1977). Impurity diffusion in amorphous silicon and its implications for solar cells. *RCA Review*, 38, 211.

Chapin, D.M., Fuller, C.S., Pearson, G.I. (1954). A new silicon p-n junction photocell for converting solar radiation into electric power. *Journal of Applied Physics*, 25(5), 676–677.

Chapin, D.M., Fuller, C.S., Pearson, G.I. (1977). Solar energy converting apparatus. Patent, US2780765.

Chittic, R.C., Alexander, J.H., Sterling, H.F. (1969). The preparation and properties of amorphous silicon. *Journal of Electrochemical Society*, 116, 77.

Coletti, G., Kvande, R., Mihailetchi, V.D. (2008). The effect of iron. Contamination in multicrystaline silicon ingots for solar cells. *Journal of Applied Physics*, 104(104913), 1–11.

Cusano, D. (1963). CdTe solar cells and photovoltaic heterojunctions in II VI compounds. *Solid State Electronics*, 6(3), 217–218.

Dastghelb-Shirazi, A., Steyer, M., Micard, G., Wagner, H., Altermatt, P.P., Hahn, G. (2013). Relationships between diffusion parameters and phosphorus precipitation during the POCl₃ diffusion process. *Energy Procedia*, 38, 254–262.

Davis, J.R., Rohatgi, A., Hopkins, R.H., Blais, P.D., Rai-Choudhury, P., McCormick, J.R., Mollenkopf, H.C. (1980). Impurities in silicon solar cells. *IEEE Transactions on Electron Devices*, 27(4), 677–687.

Fraunhofer ISE (2016). Institute for solar technology. Freiburg [Online]. Available at: www.ise-fraunhofer.de.

Fritts, C.E. (1883). On a new form of selenium photocell. *American Journal of Science*, 26, 465.

Fuller, C.S. (1962). Method of forming semiconductive bodies. Patent, US3015590.

Goetzberger, A. and Hebling, C. (2000). Photovoltaic materials, past, present and future. *Solar Energy Materials and Solar Cells*, 62, 1–19.

Gray, J.L. (2003). The physics of solar cells. In *Handbook of Photovoltaic Science and Engineering*, Luque, A. and Hegedus, S. (eds). Wiley, New York.

Hittorf, J.W. (1851). Ueber das elektrische leitungvermögen des schwefelsibers. *Ann. Phys. Lpz.*, 84, 1–28.

Honsberg, C. and Bowden, S. (n.d.). Absorption of light [Online]. Available at: http://www.pveducation.org/pvedrom.

Istratov, A.A., Buonassisi, T., Pickett, M.D., Heuer, M., Weber, E.R. (2006). Control of metal impurities in "dirty" multicrystalline silicon for solar cells. *Materials Science and Engineering*, B134(2/3), 282–286.

Jenny, D.A., Loferski, J.J., Rappaport, P. (1956). Photovoltaic effect in GaAs p–n junctions and solar energy. *Conversion Physical Review*, 101(1208).

Koch, W., Endrös, A. L., Franke, D., Haßler, C., Kalejs, J.P., Möller, H. J. (2003). Bulk crystal growth and wafering for PV. In *Handbook of Photovoltaic Science and Engineering*, Luque, A. and Hegedus, S. (eds). John Wiley and Sons, Chichester.

Kodes, J. (1971). Photovoltaic effect in selenium photocells. *Physica Statu Solidi*, 5(2), K87.

Loferski, J.J. (1956). Theoretical considerations governing the choice of the optimum semiconductor or photovoltaic solar energy conversion. *Journal Applied Physics*, 27(7), 777–784.

Ohl, R.S. (1946). Light sensitive electric device. Patent, US2402662.

Ohl, R.S. (1948). Light-sensitive device including silicon. Patent, US2443542.

Prince, M.B. (1955). Silicon solar energy converters. *Journal Applied Physics*, 26(5), 534–540.

REC Silicon (Renewable Energy Corporation) (2011). Benefit of FBR granular polysilicon [Online]. Available at: www.recgroup.com.

Reynolds, D., Leies, G., Antes, L.L., Marburger, R.E. (1954). Potovoltaic effect in cadmium. *Physical Review*, 96(2), 533.

Siemens, W. (1875). On the influence of light upon the conductivity of crystalline selenium. *Philosophical Magazine*, 50(11), 416.

Smith, W. (1873). Sélénium. *Nature*, 20(7), 303.

Smits, F.M. (1976). History of silicon solar cells. *IEEE Transactions on Electron Devices*, ED-23(7), 640–642.

Søiland, A.K., Odden, J.O, Sandberg, B., Friestad, K., Håkedal, J., Enebakk, E., Braathen, S. (2012). Solar silicon from a metallurgical route by Elkem solar – A viable alternative to virgin polysilicon. *6th International Workshop on Crystalline Silicon for Solar Cells*, 8–11 October, Aix-les-Bains.

Spear, W.E. and Le Comber, P.G. (1975). Amorphous semiconductors. *Solid State Commun.*, 17, 9.

Sze, S.M. (2002). *Semiconductor Devices*, 2nd edition. Wiley, New York.

Warren, M. (2016). Transistor museum donation [Online]. Available at: http://www.transistormuseum.com.

Digital Photographic Sensors

Charge-coupled device (CCD) and complementary metal-oxide-semiconductor sensor active-pixel sensor (CMOS-APS) digital photographic sensors have replaced photographic films.

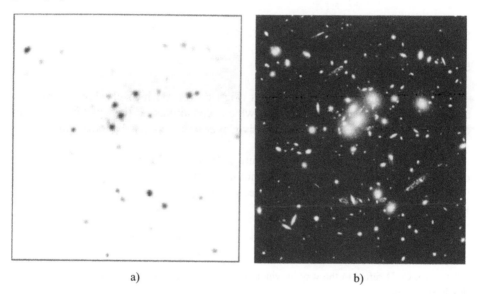

a)   b)

Figure 5.1. *(a) A silver photo of the sky; (b) a CCD image of the same portion of the sky (Smith 2009). For a color version of this figure, see www.iste.co.uk/vignes/silicon2.zip*

Continuous development from the 1980s to the present day has led to the manufacture of 100 million-pixel sensors with quantum efficiencies close to 100%, capable of detecting a single photon, and with sensitivities ranging from X-rays to infrared.

The invention of CCD photoelectric sensors in 1970 and CMOS-APS sensors in 1993 transformed photography into a digital medium. This revolutionized astronomy and medical imaging. They are also used in fax machines and scanners. Other applications for CMOS-APS sensors include webcams, high-frame-rate cameras, endoscopic cameras and cameras integrated into cell phones.

5.1. Introduction

CCD and CMOS-APS sensors are integrated circuits consisting of an array of pixels, the basic component of which is a MOS capacitor, in which the electrons generated by the photoelectric effect during the exposure time are stored, and a circuit consisting of a set of transistors converting each packet of electrons into a voltage that is digitized and transmitted for storage in a memory.

5.2. CCD sensors

5.2.1. Features

A CCD sensor[1] is an array of pixels (Figure 5.2). Each pixel (short for picture element) is a photoelectric receiver which constitutes a tiny picture element equivalent to the grain of the photosensitive layer of silver-based photographic film.

Each pixel is a MOS capacitor (Figure 5.3) whose silicon P substrate absorbs, during the exposure time, a packet of photons, which generate a packet of electrons (generation of electron–hole pairs) (see Volume 1, Chapter 1, section 1.2.5 and Figure 1.8)[2].

---

1 The term CCD refers to the way in which the information is read, not to the generation of this information.

2 The same mechanism is used to form the image on photographic film: the film consists of a plastic support film covered with an emulsion: a layer of gelatin in which silver bromide (AgBr) crystals are mixed. When exposed to light, photons impinge on the film. Each photon absorbed releases an electron from the Br⁻ anion. This electron is captured by an Ag⁺ ion, which is reduced to a silver atom. For each AgBr crystal, depending on the light intensity of the part of the subject it describes, from 0 to around 10 Ag atoms are formed. These atoms tend to stick together to form an aggregate (see "analog photography", on Wikipedia). After development, they form the grains of the photo.

At the end of the exposure time, these packets of electrons are transferred, pixel by pixel, along a column of pixels (Figure 5.2) (transferred as if emptying the buckets of a column into one another), then along a horizontal line of capacitors, making up a serial register. This register is connected to a readout circuit, which converts each charge packet into an analog voltage signal proportional to the charge packet (Q = CV). This voltage is amplified, and then digitized by an analog-to-digital converter and transmitted in series to be stored in a memory (Figure 5.6).

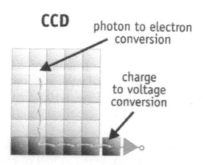

Figure 5.2. *Pixel array of a CCD and charge transfer from one pixel to its neighbor, then into the serial (horizontal) register (Litwiller 2005). For a color version of this figure, see www.iste.co.uk/vignes/silicon2.zip*

While the resolution (pixel size) of the first detectors was 20–30 μm (for fine-grain film, the average size of a silver grain is around 20 μm), in 2007 it reached less than 7 μm for CCD cameras.

Standard astronomical CCD sensors are made up of 2k × 4k pixels 15 × 15 μm, making up a 30 × 60 mm chip, offering excellent spatial resolution.

5.2.2. CCD pixel structure and quantum efficiency

The basic component is a MOS capacitor (Figure 5.3) (shown in Volume 1, Chapter 6, section 6.1.2.1, Figure 6.2) with a silicon P substrate and polysilicon front electrode separated by an oxide layer. The capture of incident photons and the generation of electron–hole pairs by photon absorption occur in the silicon substrate P. Electrons are stored in the inverted channel (see Chapter 4, section 4.5.2).

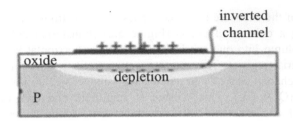

Figure 5.3. *MOS capacitance: the electrons created by the photons in the P-base are stored in the inverted channel (Kuphaldt 2009). For a color version of this figure, see www.iste.co.uk/vignes/silicon2.zip*

Silicon, in the visible range, in terms of quantum efficiency: photon–electron conversion efficiency is the best semiconductor material. Because of its energy gap of 1.12 eV, it absorbs wavelengths up to 1,100 nm, that is, into the infrared (see Chapter 4, section 4.4.3, Figure 4.9).

The advantages of the CCD over photographic films: the spectral sensitivity of the CCD extends from 300 to 1,000 nm, whereas that of photographic film extends from 350 to 700 nm. Quantum efficiency, that is, the fraction of photons actually collected, for silver film is only 1–4%, whereas it varies from 40% to over 80% depending on the pixel structure and wavelength for the CCD sensor (Figure 5.4(c)).

5.2.2.1. *Thick CCD pixel and thin CCD pixel*

Two types of CCD pixel have been developed: thick and front-illuminated, thinned and back-illuminated to increase quantum efficiency (Figures 5.4(a) and (b)).

The thick CCD pixel (150–500 μm) is illuminated from the front through the polysilicon front electrode (gate). Quantum efficiency is low, due to the absorption of energetic photons by the front electrode. It reaches 40% for the 700 nm wavelength. It is very low in the blue (black curve, Figure 5.4(c)).

The thin CCD pixel (< 50 μm) is illuminated from behind, incident photons directly strike the photosensitive wafer. Quantum efficiency reaches 0.9 for an 800 nm photon (red). It is much more difficult to manufacture than the front-illuminated CCD (red curve, Figure 5.4(c)).

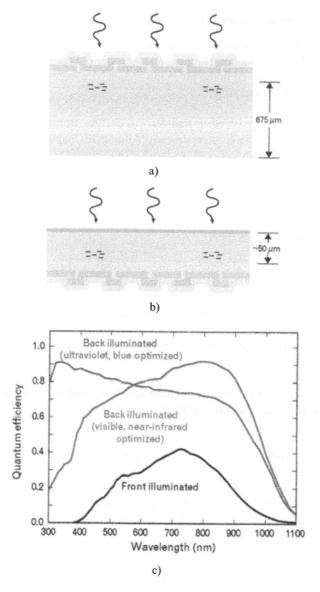

Figure 5.4. *(a) Thick pixel illuminated from the front; (b) thin pixel illuminated from the back; (c) quantum efficiencies of front-illuminated CCDs and back-illuminated CCDs optimized for short and long wavelengths (Burke et al. 2007, pp. 395 and 397). For a color version of this figure, see www.iste.co.uk/vignes/silicon2.zip*

5.2.2.2. *CCD pixel with surface channel and CCD pixel with buried channel*

In a CCD pixel with a surface channel, electrons are stored in the P-base inverted channel of an MOS capacitor (Figure 5.3) biased to create an n-type inverted channel. The electrons stored in the inversion channel are in contact with the Si/SiO$_2$ interface of the MOS capacitor.

Given the problems involved in transferring charges from pixel to pixel due to charge trapping at the Si/SiO$_2$ interface (see Volume 1, Chapter 3, section 3.1.2.1), in order to move electrons away from the Si/SiO$_2$ interface, an n-buried channel CCD pixel was developed (Figure 5.6). It consists of a layer of silicon N between the layer of SiO$_2$ and the P base. The pixel is made up of a PN diode coupled to a MOS capacitor. The electrons created in the P base are stored in the silicon N layer, called "buried channel". Electrons are stored "deep down", and the charge transfer from one pixel to the next is total.

For a pixel with a surface channel, the quantum yield reaches 0.99, and for a pixel with a buried channel, the quantum efficiency reaches 0.999999.

The CCD pixel capacity is the maximum number of electrons generated by incident photons and stored either in the inversion channel or in the silicon layer N depending on the type of pixel. With a 15×15 μm pixel, a capacity of 200–250 ke is achieved.

Parasitic charges are created by thermal agitation (dark current). To reduce the level of dark current, the CCD in astronomy is cooled down.

5.2.3. *CCD cell structure and charge transfer*

A CCD cell actually consists of three closely related MOS capacitors: a collection capacitor, the pixel itself (phase 1), whose electrode is raised to a positive potential to create a potential well, and two transfer capacitors, known as barriers.

The three capacitors are connected to clock signal lines. The structure is shown in Figure 5.5(a), and the transfer of charges from the collection capacitor where the electrons are photogenerated through the two transfer capacitors is shown in Figure 5.5(b). A judicious succession of polarization of these capacitors allows charges to pass from one pixel to the next.

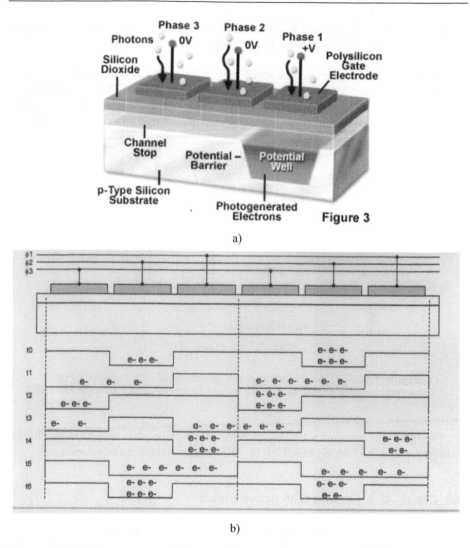

Figure 5.5. *(a) Structure of a surface-channel CCD cell (Howell 2003); (b) charge transfer from near to near within a three-capacitor CCD cell and from one cell to another (Boulade 2007). For a color version of this figure, see www.iste.co.uk/vignes/ silicon2.zip*

5.2.4. Reading circuit

This is the electronic circuit consisting of a set of transistors converting the charge of a pixel into an analog voltage proportional to the stored charge (Q = CV)

(Figure 5.6), which is then amplified and digitized by an analog-to-digital converter and transmitted in series for storage in a memory.

The electron packet leaving the serial register falls into a "summation well" n^+. The transfer is controlled by a first MOSFET transistor (reset FET). The charge eN on a pixel induces a voltage V, which in turn activates a second MOSFET (sense FET) to amplify the signal, which is then transferred to external electronics for digitization.

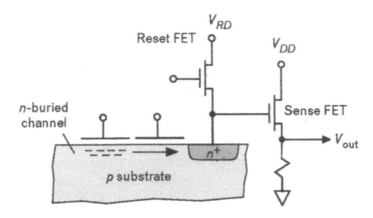

Figure 5.6. *CCD pixel reading circuit (Burke et al. 2007, p. 399). For a color version of this figure, see www.iste.co.uk/vignes/silicon2.zip*

The main limitation of a CCD sensor lies in its principle: close-to-close charge transfer reading. To remedy this, CMOS-APS sensors have been developed.

5.3. CMOS-APS sensors with active pixels

The CMOS-APS sensor is derived from the CCD. Each active pixel (APS) consists of a MOS capacitor and a set of transistors for in situ conversion of the accumulated charge into a voltage (Q = CV) and voltage amplification (Figure 5.7). Each pixel contains its own readout electronics, so there are no more transfer problems (Figure 5.8).

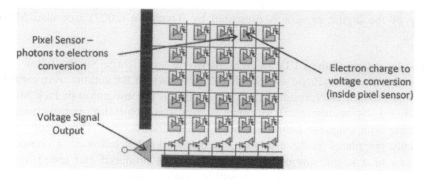

Figure 5.7. *Active pixel matrix (APS) of a CMOS-APS sensor (Fossum 1997; Litwiller 2005). For a color version of this figure, see www.iste.co.uk/vignes/silicon2.zip*

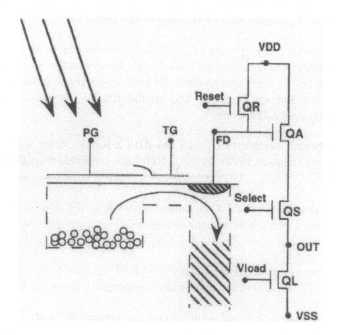

Figure 5.8. *Structure of an active pixel (APS) of a CMOS-APS sensor (Dickinson and Fossum 1995)*

5.4. History

The history of the discovery and development of digital image sensors is described in detail in *Scientific Charge-Coupled Devices* (Janesick 2001). The history of the digital camera is presented by Trenholm (2007) (see also Maitre (2017)).

The invention of the CCD, by Willard Boyle and George Smith of Bell Labs (yet another) in 1970 (Boyle and Smith 1970), for which the authors were awarded the Nobel Prize in 2009, resulted from the desire (at the instigation of Jack Morton from Bell Labs management) to create a semi-conductor shift register memory competing with magnetic bubble memory, itself serial access[3]. The invention was reportedly completed in the afternoon. In the weeks that followed, a component consisting of a single row of MOS capacitors was produced and tested in their publications, the authors indicate possible applications: shift register, imaging device and display devices.

In view of the problems posed by the transfer of charge from one MOS capacitor to its neighbor due to charge traps at the Si/SiO_2 interface, the authors invented the buried-channel CCD, introduced in 1973 (Boyle and Smith 1974).

One factor in the development of CCD sensors was the need to replace the cameras using *vidicon* tube sensors with solid-state sensors on the *Surveyor*, *Ranger*, *Mariner*, *Viking* and *Voyager* space probes for solar system exploration, designed by the Jet Propulsion Laboratory (JPL), and on the JPL LST space telescope, later called the *Hubble Space Telescope*.

In 1972, Michael Tompsett of Bell Labs filed a patent on the application of a CCD to imaging (Tompsett 1978). In 1972, Bell Labs presented the CCD to the LST and JPL teams.

But, once again, Bell Labs did not pursue research on this component, as parent company Western Electric was not authorized to sell such components outside of its industrial domain.

The developments pursued by the JPL involved three manufacturers: Fairchild Semiconductor, RCA Corporation and Texas Instruments (Janesick 2001).

Fairchild, as soon as it heard of the invention, sensing the interest of the CCD concept, "poached" Gil Amelio from Bell Labs, who had helped develop the first CCD, Another Bell Labs defector, James Early, had joined Fairchild in 1969. He

---

3 Magnetic bubble memories were invented by Bell Labs in September 1967. The main inventor was Andrew Bobeck. By the mid-1970s, virtually every major electronics company had a team working on bubble memories. Texas Instruments marketed the first equipment incorporating bubble memories in 1977. Bubble memories disappeared entirely from the market in the late 1980s, to be replaced by faster, higher capacity and more economical hard disks.

was the driving force behind the buried-channel CCD. However, the architecture of the array was such that a large part of the front surface was not illuminated, resulting in low quantum efficiency. The first commercial CCDs manufactured by Fairchild appeared in 1974. In 1974, the first astronomic image was taken with a 100 × 100 pixels (Fairchild CCD).

Meanwhile, the RCA company was developing a thinned, backside illuminated CCD whose quantum efficiency was much higher than the CCD developed by Fairchild. In 1979, RCA's Dick Savoye developed the largest CCD, 512 × 320 pixels, for television applications. Unfortunately, RCA's CCD was based on surface-channel pixels rather than buried-channel pixels. What is more, its spatial resolution was unsatisfactory for scientific applications. It was far less powerful than the CCDs manufactured by Fairchild.

The JPL, under NASA contract, teamed up with Texas Instruments to develop a "scientific" sensor. The collaboration between JPL and TI lasted some 10 years and resulted in a technology used today: a CCD based on full-frame, backside-illumination buried channel technology, thinned (< 50 μm) and illuminated from the rear. In 1975, JPL announced "a 100×160 backside-illuminated TI CCD", and in 1976 a 400×400 TI CCD sensor appeared.

In 1990, the first large-format 2k × 2k CCDs appeared.

Continuous development from the 1980s to the present day has led to the realization of 100 million-pixel sensors, with quantum efficiencies close to 100%, capable of detecting a single photon, and with sensitivities ranging from X-rays to infrared.

As astronomical sources are generally very faint, astronomical sensors can have exposure times ranging from a fraction of a second to a few hours.

Current manufacturers of astronomical sensors include Texas Instruments, Fairchild and Tektronix. In 1975, Steven Sasson of Kodak built a prototype digital camera using a CCD manufactured by Fairchild (Lloyds and Sasson 1978). The first commercial CCD camera was developed by Fairchild in 1976 (MV-101) and used to monitor products manufactured by Procter & Gamble. In 1977, Konica launched the first C35-AF autofocus camera.

The real start (the filmless age) of digital photography is considered to date from August 25, 1981, when Sony released the Mavica (Magnetic Video Camera), using a 570 × 490 pixels CCD on a 10 × 12 mm chip.

The year 1993 saw the invention of the CMOS-APS sensor by Eric Fossum (Fossum 1993). With this new technology, APS sensors became the commercial successors to CCD sensors. Between 1993 and 1995, several prototypes were developed for a number of applications by the JPL using these new APS imagers. These included webcams, high-capacity cameras, digital radiography, medical X-ray imaging, endoscopic cameras and cameras integrated into cell phones.

5.5. References

Boulade, O. (2007) Imagerie CCD en astronomie. École IN2P3 du détecteur à la mesure. Saclay, 1–122 [Online]. Available at: http://formation.in2p3.fr.

Boyle, W.S. and Smith, G.E. (1970). Charge coupled semiconductor device. *Bell System Technical Journal, Briefs*, 49(4), 587–593.

Boyle, W.S. and Smith, G.E. (1974). Buried channel charge coupled devices. Patent, US3792322.

Burke, B.E., Gregory, J.A., Cooper, M., Loomis, A.H., Young, D.J., Lind, T.A., Doherty, P., Daniels, P., Landers, D.J., Ciampi, J. et al. (2007). CCD image development for astronomy. *Lincoln Laboratory Journal*, 16(2), 393–410.

Dickinson, A. and Fossum, E.R. (1995). Standard CMOS active pixel image sensors for multimedia applications. In *16th Conference on Advanced Research*, 27–29 March. VLSI, Chapel Hill.

Fossum, E.R. (1993). Active pixel sensors: Are CCD's dinosaurs? 1–13 [Online]. Available at: https://www.spiedigitallibrary.org.

Fossum, E.R. (1997). CMOS active pixel sensor (APS) technology for multimedia image capture. In *Multimedia Technology and Applications Conference*, 23–25 March. University of California, Irvine.

Howell, S.B. (2003). Basics of charge-coupled devices. *NOAO*, 1–27 [Online]. Available at: https://noirlab.edu.

Janesick, J.R. (2001). *Scientific Charge-Coupled Devices*. SPIE Press, Bellingham.

Kuphaldt, T.R. (2009). Lessons in electric circuits, volume 3 [Online]. Available at: openbookproject.net/electriccircuits.

Litwiler, D. (2005). CCD vs CMOS. Maturing technologies, maturing markets [Online]. Available at: www.dalsa.com.

Lloyds, G.A. and Sasson, S.J. (1978). Electronic still camera. Patent, US4131919.

Maitre, H. (2017). *From Photon to Pixel: The Digital Camera Handbook*, 2nd edition. ISTE Ltd, London, and John Wiley & Sons, New York.

Smith, G.E. (2009). The invention and early history of the CCD. Nobel Lecture.

Tompsett, M. (1978). Charge transfer imaging device. Patent, US4085456.

Trenholm, R. (2007). Photos: The history of the digital camera. *C/NET archives* [Online]. Available at: https://www.cnet.com.

6

Microelectromechanical Systems

Silicon's exceptional mechanical properties, combined with its electrical properties, make it the material of choice for precision micromechanics and electromechanical microsystems (MEMS).

The history of MEMS began with the discovery of the piezoresistive effect of silicon and germanium in 1954.

A MEMS is a microdevice that acts as a sensor (converting a mechanical stress into an electrical signal), an actuator (converting an electrical signal into a mechanical action) or a resonator (imposing a resonant frequency on the structure).

A MEMS is a mechanical structure of very small dimensions (a few square millimeters), produced in a single-crystal silicon substrate using microelectronics technologies, with at least one micrometric movable element.

A MEMS "sensor" or "actuator" is a basic component of what we call intelligent objects, since it provides us with information about our environment and vice versa.

The development of these microsystems has been facilitated by the availability of mastered microelectronics manufacturing technologies, giving MEMS high precision and reproducibility. But development has been relatively slow, due to the need to develop specific miniaturization technologies, particularly for structures with moving parts.

This chapter presents:

– a brief history;

– MEMS operating principles (sensors and actuators);

– some examples of MEMS;

– MEMS manufacturing technologies.

6.1. Parts for precision mechanics

In the watchmaking industry, which until recently has used exclusively metallic mechanical parts, silicon parts have been present in high-precision mechanical watches since 2006 (Figures 6.1 and 6.2) (UFC 2011). These parts are manufactured from single-crystal silicon wafers.

The properties of single-crystal silicon and physicochemical micromachining techniques (presented in section 6.5) enable us to manufacture high-precision micromechanical parts to micrometer tolerances.

What is more, for micrometric parts up to 1-mm thick, silicon retains all the mechanical properties that characterize it and differentiate it from conventional metals.

Insensitive to magnetic fields, a material of spirals, highly resistant to corrosion and wear, lighter and harder than steel, silicon reduces inertia and enables lubrication-free operation.

Polished to near-nanometric fineness, it produces parts with excellent surface quality.

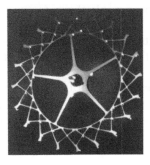

Figure 6.1. *Silicon escapement wheels (Breguet watches) (www.breguet.com).*
For a color version of this figure, see www.iste.co.uk/vignes/silicon2.zip

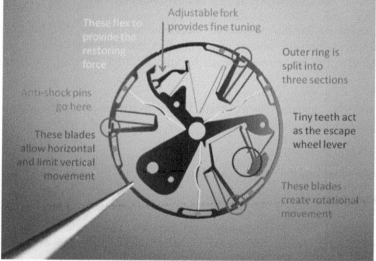

Figure 6.2. *New two-component oscillator (purple) in monocrystalline silicon (replacing the balance-spring duo) (Zenith Defy Lab watch) (LVMH). For a color version of this figure, see www.iste.co.uk/vignes/silicon2.zip*

6.2. MEMS: sensors and actuators

6.2.1. *Presentation*

A MEMS is a microdevice that acts as a sensor, actuator or resonator (imposing a resonant frequency on the structure) (Wang 2006).

A MEMS is a mechanical structure of very small dimensions (a few square millimeters), produced in a single-crystal silicon substrate by microelectronics technologies with at least one movable micrometric element. By extension, microdevices machined into a silicon substrate by microelectronics technologies, such as microchannels etched into a silicon substrate, used as mixers or tubular microreactors in chemistry and biochemistry, are considered MEMS microfluidics.

The MEMS "sensor" is a system (structure) comprising a fixed and a moving element (membrane, embedded beam, embedded bridge, seismic mass, etc.), whose reaction (displacement of the mass or deformation of a membrane, vibration of a beam) to a physical stimulus (force, pressure, acceleration, etc.) is converted into an electrical signal.

Examples of sensors that have seen considerable recent development include accelerometers and gyroscopes, as well as micromicrophones.

The sensor is combined with an electronic component that transforms the electrical signal from the sensor, which is often unusable because it is too weak, distorted and noisy, into a useful signal by amplification, filtering, modulation and so on.

The MEMS "actuator" is a system that converts an electrical signal into a mechanical action. Examples include:

– inkjet printer heads;

– micropumps and microvalves for fluidics circuits;

– micromotors.

The MEMS "resonator" combines an actuator to set the moving structure into vibration, whose vibration frequency depends on the parameter to be quantified, and a sensor to measure this frequency (e.g. resonant beam sensor for species detection; Figure 6.13). An important application, in telecommunications, is an electromechanical radio-frequency filter: FBAR (film bulk acoustic resonator) (Figure 6.10). On the other hand, time-based generator resonators have not succeeded in dethroning quartz in clocks.

A MEMS "sensor" or "actuator" is an essential component of what are known as "intelligent objects", since it provides us with information linked to our environment, and vice versa. A series of technological breakthroughs and industrial gambles have helped to explode a market that continues to evolve (Vigna 2013).

An example includes the electronic trajectory control system that requires the use of inertial sensors.

6.2.2. *History*

The history of MEMS began with the discovery of the piezoresistivity of silicon and germanium, in 1954, by Smith (1954), a Bell Labs researcher (yet another Bell Labs discovery). The fabrication of a silicon strain gauge, integrated into the moving part of a structure (Figure 6.1), was carried out by Pfann and Thurston in 1961 (Pfann and Thurston 1961; Pfann 1962), and this sensor was first implemented on single-crystal silicon membranes by Tufte in 1962 (Tufte et al. 1962; Tufte and Stelzer 1963).

The first developments in MEMS began in the early 1970s.

The first publications describing piezoresistive and capacitive pressure sensors appeared in 1973. Pressure transducers were commercialized in the early 1980s.

The first MEMS microphone with silicon diaphragm and piezoelectric gauge ZnO layer was developed by Royer in 1983 (Royer et al. 1983) and with an AlN layer in 1988 (Frantz 1988). The first acoustic MEMS devices with a PZT gauge $(Pb(Zr_xTi_{1-x})O_3)$ were produced in the 1990s. A microphone with a highly boron-doped P silicon diaphragm and a polysilicon piezoresistive gauge was developed by Schellin and Hess in 1992 (Schellin and Hess 1992). The first capacitive micromachined silicon microphone was developed in 1983 (Hohm and Sessler 1983), and the first capacitive microphones were introduced to the market in 2003 (Adell 2003) (see Figure 6.14).

The first publications describing two accelerometers appeared in 1979: one piezoresistive (Roylance and Angell 1979) and the other capacitive (Chen et al. 1982). In 1984, a patent (Marcillat 1985) was filed describing a capacitive lateral accelerometer with interdigital combs (Figure 6.16) (a device that was adopted by virtually everyone in the field), and in 1989 the first SFENA accelerometer sensor went into production. The electrostatic actuator with interdigitated combs was to dominate the world of actuators (Figures 6.6–6.8).

But it is in the 21st century that MEMS really took off. The revolution took place in the 2000s and was triggered by a Nintendo video game console. It uses a system capable of detecting position, orientation and movement in space using MEMS accelerometers and gyroscopes (according to Benedetto Vigna of STMicroelectronics, one of the leaders) (Vigna 2013).

In less than a decade, MEMS have become fundamental components of most of the high-tech products that surround us. We can no longer do without them. Today, a smartphone contains up to 10 MEMS (accelerometers microphones, acoustic sensors to suppress background noise, etc.). A smartphone can accurately estimate its location in space by coupling a compass, an accelerometer and a gyroscope. A gyroscope provides optical image stabilization for the camera.

According to Benedetto Vigna, "This sector is engaged in a truly exceptional dynamic of innovation: all kinds of applications are emerging, and worlds are opening up to our technologies. For example, the automotive industry, which was one of the first users of MEMS with airbags, has begun to turn its attention to other uses, such as guidance systems based on acoustic transducers (sonar) for turning corners, ESP devices, electronic trajectory control systems using vibrating gyroscopes. After finding their way into our cars and smartphones, inertial MEMS sensors are now finding their way into all connected objects. This increasingly high-performance technology is also winning over manufacturers for navigation applications in avionics, space and self-driving vehicles" (Vigna 2013).

The medical sector, and beyond that the whole nebula of well-being technologies, are also major users: all the instruments that tomorrow will enable you to take your blood pressure while walking or measure your diabetes levels at home use MEMS.

6.2.3. Silicon: the material of MEMS

The mechanical structure of MEMS consists of a fixed part etched on a substrate, a wafer made of monocrystalline silicon and a moving part (also made of silicon, very often polycrystalline) produced by machining or by deposition processes such as LPCVD (low-pressure chemical vapor deposition).

The use of microsystems has grown as a result of their small size, combined with high sensitivity, linked to the mechanical and electrical properties of silicon, and high precision and reproducibility, linked to manufacturing technologies.

The size of a MEMS is of the order of a square millimeter, while the moving parts of its structure are of the order of a micron. Thanks to their high level of miniaturization, MEMS can be integrated into a wide range of systems.

The factors that have contributed to silicon's dominance are as follows:

– silicon's exceptional mechanical properties, conferring a very great reliability to the structure and its reactions to mechanical stresses (see Box 6.1);

– suitable electrical properties;

– electrical conductivity that can be controlled by doping;

– very high piezoresistivity;

– manufacturing technologies mastered through microelectronic technologies, conferring high precision and reproducibility to MEMS, as well as low costs through massive collective processes. The production of these microsystems has necessitated the development of new processes (volume micromachining and surface micromachining presented in section 6.5), which explains the long development times involved. In addition, as the moving part is produced by deposition processes such as LPCVD, which induce highly variable residual stresses in these thin films, modifying their mechanical properties, specific heat treatments had to be incorporated into the manufacturing process;

– the possibility of integration with integrated circuits.

The properties are as follows:

– reproducible material of very high crystallographic quality;

– near-perfect Hooke material: reproducible elastic deformations, no hysteresis, which means that when bent, it undergoes virtually no hysteresis and consequently no energy dissipation or fatigue;

– Young's modulus close to that of steel;

– lighter than steel, density close to that of aluminum;

– no plastic deformation;

– fracture stress twice that of steel;

– very good resistance to mechanical fatigue (good ageing);

– thermal conductivity ~50% higher than steel;

– thermal expansion: 1/5 that of steel.

Box 6.1. *Mechanical properties of single-crystal silicon*

Miniaturization has both positive and negative consequences as a result of the laws of scale.

For an embedded beam (lever, cantilever beam), the inertial forces, for a force F applied to its free end, vary as the seismic mass m: that is, as s^3, the mechanical strength (stiffness k) varies only as s and the deflection varies as $1/s^2$, where s is the miniaturization factor.

Stiffness:

$$k = E \, wt^3/l^3$$

where w is the width, t the thickness and l the length of the beam.

Arrow or displacement:

$$h = 4Fl^3/Ewt^3 = (m/k) \times (\gamma)$$

where γ is the acceleration.

A positive consequence of this characteristic is that such a structure can withstand enormous accelerations without breaking. But a negative consequence is that the displacement of the end of an embedded beam (the deflection) varies as $1/s^2$. It is very small, requiring a highly sensitive displacement detection system.

In the case of an embedded resonant beam, the micrometric dimension of the mobile structure results in very high resonance frequencies, making it possible to detect very small mass variations (gravimetric sensor) (Figure 6.13):

$$f \text{ (or } v) = 1/2\pi \, (m/k)^{1/2}$$

6.3. The physical basis of MEMS operation

6.3.1. *Sensors*

The operation of a MEMS sensor is based on the conversion of a mechanical signal into an electrical signal, and vice versa for an actuator.

The MEMS "sensor" is a system (structure) comprising a fixed element and a deformable or mobile element (embedded beam, embedded bridge, membrane, seismic mass, etc.), whose reaction (displacement of the mass, deformation of a membrane, vibration of a beam) to a physical stimulus (force, pressure, acceleration, etc.) is converted into an electrical signal.

Depending on the MEMS structure and the type of conversion of the mechanical stress (a force, pressure, acceleration, rotation, deformation or displacement of the moving part), its conversion into an electrical signal can be achieved by piezoresistive or piezoelectric gauges measuring a deformation of the moving part of the MEMS, or by capacitive detection measuring the displacement of the moving part of the MEMS.

We present these types of structures and conversions used on accelerometers.

6.3.1.1. *Piezoresistive detection*

A schematic diagram of a piezoresistive gauge accelerometer is shown in Figure 6.3. It consists of a seismic mass attached to the end of a beam.

Piezoresistive gauge

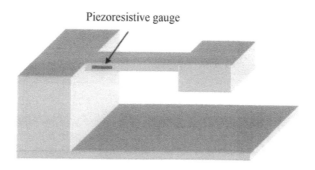

Figure 6.3. *Piezoresistive gauge accelerometer (Chaehoi 2005, p. 18)*

Piezoresistivity is the variation of the electrical resistivity of a material caused by mechanical stress. A stress applied to a silicon rod will modify its resistance for two reasons: its geometric variation, but also the variation of the resistivity of the material. The piezoresistance of a metal sensor is due solely to the change in geometry caused by mechanical stress (K = 1 + 2 ν). Piezoresistive gauges in doped silicon are produced by diffusion in the moving part of the structure (beam or vibrating diaphragm).

6.3.1.2. *Piezoelectric detection*

Figure 6.4 shows an example of a piezoelectric accelerometer.

Piezoelectricity is the property of certain bodies to become electrically polarized under the action of mechanical stress (i.e. to present a potential at their terminals according to the variation in stress), and conversely to deform elastically when a voltage is applied. Gauges consist of piezoelectric thin films (ZnO, AlN, PZT) deposited on the moving part under stress. Piezoelectric detection is used in high-frequency applications.

Piezoelectric gauge MEMS have been the subject of numerous developments, but the problems posed by the deposition of piezoelectric thin films (ZnO, AlN, PZT) on the moving part of the mechanical structures are holding back their commercialization.

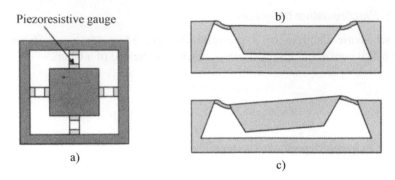

Piezoresistive gauge

a)

b)

c)

Figure 6.4. *Three-axis accelerometer with piezoelectric detection:*
(a) top view of structure; (b) vertical acceleration; (c) lateral
acceleration (Scheeper et al. 1996; Chaehoi 2005, p. 20)

6.3.1.3. Capacitive detection

Figure 6.5 shows the structure and operation of the capacitive accelerometer. A capacitive sensor is a displacement sensor consisting of two (flat) electrodes. One is fixed, while the other, the seismic mass, is mobile and attached to a spring, forming a mass/spring system. Acceleration generates a force applied to the movable electrode, producing a displacement of this electrode relative to the fixed one. The relative displacement of two plates induces a variation in capacitance:

$$C = \varepsilon \, S/d$$

with:

– $S = lw$: surface area (w width, l length);

– d: distance;

– ε: the dielectric permeability of the substance placed between two plates.

Such a sensor, based on the measurement of the variation in the distance "d" between the two electrodes, enables extremely sensitive measurements of (dynamic) displacements: the capacitance varying as the inverse of the distance separating the two plates, and the sensitivity varying as the inverse of the square of the distance:

$$C = C_0 \, d_0 \, /d$$

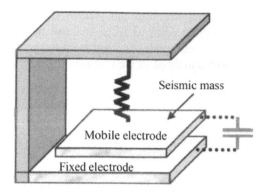

Figure 6.5. *Principle of the capacitive accelerometer (Chaehoi 2005, p. 16)*

A capacitive structure can be made entirely of silicon. Doped silicon enables the electrodes making up the moving and fixed parts of the structure to be polarized (accumulate charges on a very thin layer), making these structures comparable to capacitors. The movement of one of the armatures (electrodes), resulting in a temporal variation of the capacitance, induces a measurable displacement of charges, the motional current, *and* therefore the measurement of very small displacements, conferring very high sensitivity.

Capacitive sensors are one or more orders of magnitude more sensitive than piezoresistive sensors (Clarke 1979).

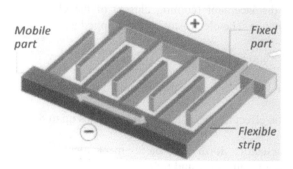

Figure 6.6. *Interdigitated combs (finger capacitors). Lateral displacements (Marcillat 1985). For a color version of this figure, see www.iste.co.uk/vignes/silicon2.zip*

The "capacitive" structures most commonly used in sensors and actuators consist of interdigitated combs (Figure 6.6). One of the combs, the moving comb, is interdependent with the moving structure and follows its movement. The flexible strip forms the spring. The other comb is fixed (relative to the substrate). Lateral displacement of the moving comb produces a variation in the distance between the lamellae of the moving comb and those of the fixed comb (Marcillat 1985).

6.3.2. *Actuators*

Piezoelectric, electromagnetic, thermal and electrostatic actuators have been developed.

Piezoelectric actuators: piezoelectric materials (ZnO, PZT) generate high forces at low displacements and very high vibration frequencies.

Electromagnetic actuation (excitation) is based on the Laplace force experienced by a conductive track placed in a magnetic field and traversed by an alternating current (Figure 6.13), which causes the beam to bend and vibrate.

The main actuator structures use the electrostatic effect. The electrostatic effect is the reciprocal of the capacitive effect.

Electrostatic actuation is based on the force created by an electrical voltage on an elementary structure consisting of two parallel flat electrodes forming a capacitor (Figure 6.5). The use of doped silicon makes the material sufficiently conductive to properly polarize the actuator armatures.

A voltage V applied to the electrodes of the capacitor induces a force F which causes a relative displacement (coming closer) of the electrodes and therefore a variation in capacitance. The electrostatic force applied is:

$$W = 1/2CV^2 \text{ and } F_z = \delta W/\delta d = \tfrac{1}{2} V^2 (\delta C/\delta d) = \varepsilon V^2 w/2d$$

where $C = \varepsilon S/d$ and where $S = lw$ is the area and d is the spacing.

The electrostatic effect allows actuation frequencies up to the gigahertz range.

Comb-drive actuators form the basic mechanical structure acting as an electrostatic actuator (Figures 6.7 and 6.8).

The two combs move vertically toward each other, resulting in a variation in the surface area "S" common to the two electrodes, and therefore in the capacity of the combs. This mode of operation allows large displacements.

$$C = C_0 S/S_0$$

As long as the two combs remain perfectly parallel to each other, lateral attraction forces are zero. However, if the two combs are eccentric during their movements, electrostatic forces will bend the branches of the combs. If the two electrodes touch, the actuator will be destroyed by the current.

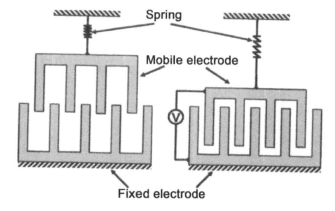

Figure 6.7. *Electrostatic actuator with interdigitated combs with vertical displacements (large displacements). For a color version of this figure, see www.iste.co.uk/vignes/silicon2.zip*

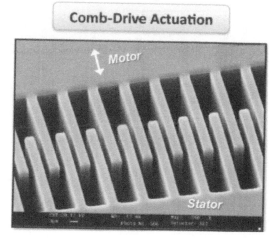

Figure 6.8. *Electrostatic actuator with interdigitated combs (DRIE etched comb-drive structure, see section 6.5.1.2) (Serrano 2013)*

6.3.3. *Microelectromechanical resonators*

A MEMS resonator, a frequency filter, consists of two interdigitated comb structures (an electrostatic actuator and a capacitive sensor) and a vibrating mass-spring structure (folded beam) imposing a resonant frequency (Figure 6.9) in the gigahertz range.

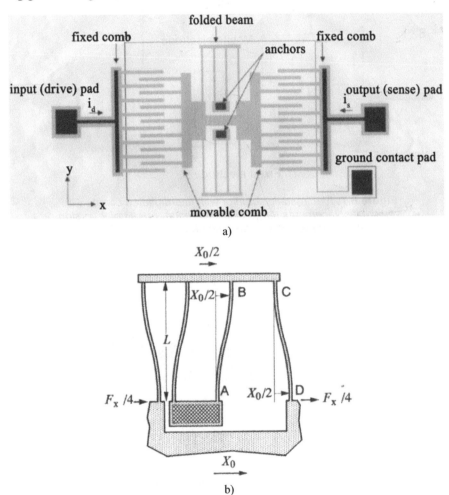

Figure 6.9. *(a) Comb-drive resonator: frequency filter; (b) folded-beam support (Tang 1990)*

A frequency-filtering MEMS resonator designed by Tang in 1990 (Tang 1990) is shown in Figures 6.9 and 6.10.

The structure on the left, consisting of two interdigitated combs is the actuator (see Figure 6.7). It produces the x-axis displacement of the central "mass-spring" structure. The folded beam (Figure 6.9(b)), 140 μm long, forms the spring. It consists of flexible lamellae (B and C) and an anchoring pot (A). The figure shows the displacement X_0 under the action of force F_x. This central structure, vibrated by the actuator, sets the resonance frequency of the overall structure. The structure on the right, made up of 2 interdigitated combs, is the sensor, whose vibration frequency is recorded capacitively.

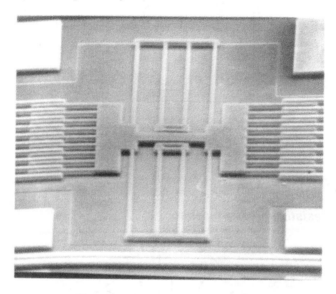

Figure 6.10. *SEM of the resonator*
(Tang 1990, p. 69)

An interdigital comb micromotor consisting of a toothed wheel driven by a beam moved by a resonant interdigital comb structure controlled by a pendulum was proposed by Tang (Figure 6.11).

In 2008, Silmach prototyped a micro-motor of this type (integrated quartz watch movement) powered by a microchip. These "watchmaking" micromotors are intended for Timex's connected watches (Silmach 2009).

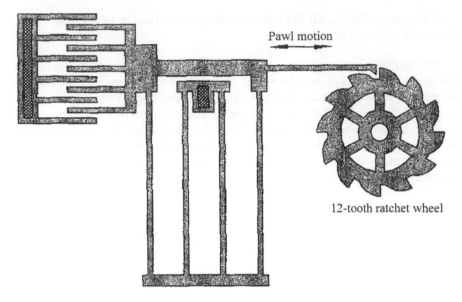

Pawl motion

12-tooth ratchet wheel

Figure 6.11. *A gear motor concept*
(Tang 1990, p. 23)

6.4. Some examples of MEMS

6.4.1. *Piezoresistive pressure sensor*

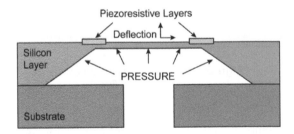

Figure 6.12. *Pressure sensor with deformable diaphragm and piezoresistive gauge in doped silicon (Prime Faraday technology watch, 2002). For a color version of this figure, see www.iste.co.uk/vignes/silicon2.zip*

The associated measurement electronics is a Wheatstone bridge, and variation in the resistivity of the piezoresistive gauge causes the output voltage to vary.

A medical sensor of this type can be placed at the end of a catheter and inserted into the veins (dimensions: $1 \times 0.7 \times 0.175$ mm).

6.4.2. *Gravimetric resonant sensor with vibrating beam*

The principle of a resonant sensor is to set the mechanical structure into vibration, the resonance frequency of which depends on the parameter to be quantified (Figure 6.13).

The mechanical element of the sensor is a microbeam (or microbridge) coated with a layer of polymer that absorbs one of the gaseous compounds produced by the chemical reaction. A piezoresistive gauge integrated into the beam detects the vibration, the frequency of which varies as a function of the beam's mass m and stiffness k, $f = 1/2\pi \, (m/k)^{1/2}$. The resonant mechanical system incorporates a magnetic actuator that acts on the structure via an external force to set it into vibration (Howe and Muller 1986).

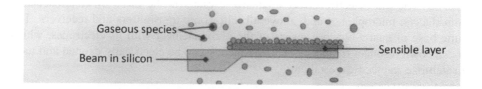

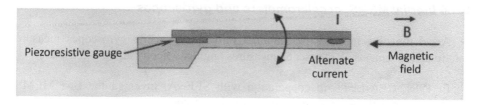

Figure 6.13. *Vibrating beam microresonant gravimetric sensor (Molinaro and Français 2015, p. 8). For a color version of this figure, see www.iste.co.uk/vignes/silicon2.zip*

6.4.3. *Microphone: capacitive sensor*

The microphone is a capacitive sensor that converts an acoustic (pressure) wave into an electrical signal (Figure 6.14). It is a "capacitor" made up of two polysilicon plates. One of the plates (the back plate) is fixed, pierced with "acoustic" holes to let the sound through, and covered with an electrode to make it conductive. The other plate, the diaphragm (diameter 750 μm), is a recessed, conductive beam. The capacitor has an extremely high impedance (gigaohm), enabling a high frequency response.

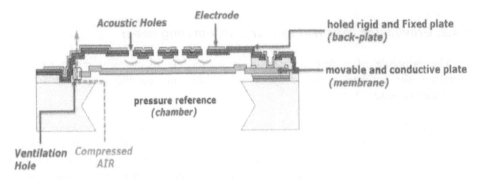

Figure 6.14. *Capacitive microphone AN4426 JRC (New Japan Radio) (STM 2017). For a color version of this figure, see www.iste.co.uk/vignes/silicon2.zip*

NOTE.– The piezoelectric effect of quartz crystals is the basis of electroacoustic transducers: microphones and, above all, ultrasonic transmitters and receivers. The time base of quartz clocks consists of the piezo element with its electrodes, which has its own resonant frequency. This frequency can be electrically excited and used to stabilize an electrical oscillator.

6.4.4 *Inertial MEMS: accelerometers and gyroscopes*

Accelerometers measure translational movements, while gyroscopes measure rotational movements. A gyroscope measures angular velocity along 1–3 axes.

In 1984, CEA-LETI, in cooperation with SFENA, invented the first MEMS accelerometer with two face-to-face interdigitated combs, and the first patent was filed by SFENA (Marcillat 1985). SFENA's first industrialization dates back to 1989. This technology has been adopted by almost all players in the field.

6.4.4.1. *The accelerometer*

The accelerometer (seismic mass/spring system) measures acceleration (change in speed) between two given instants, but not at a given instant relative to a fixed reference frame.

An accelerometer consists of a moving "seismic mass" m and a spring (stiffness constant k). An acceleration creates a force that causes the moving mass (m) to move until the force exerted by the spring (stiffness constant k) compensates for the force due to the acceleration. By measuring the variation in distance between the fixed and moving parts, we can deduce the value of the acceleration (accelerometer) or rotation (gyrometer):

Displacement of moving mass = (m/k) × acceleration

Three types of accelerometer have been developed: capacitive, piezoresistive and piezoelectric:

– Piezoelectric and piezoresistive accelerometers, where the seismic mass is positioned at the free end of a lever (flexible lamella) whose displacement is monitored by a strain gauge attached to the lever (Figure 6.3).

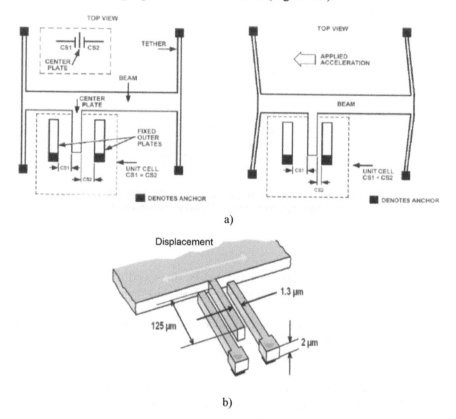

Figure 6.15. *(a) Capacitive accelerometer (Français n.d.); (b) capacitive bridge (Velay 2010). For a color version of this figure, see www.iste.co.uk/vignes/silicon2.zip*

– Capacitive accelerometers are by far the most common (Figure 6.6).

Figure 6.15 shows the schematic diagram of a capacitive accelerometer manufactured by Analog Devices, ADXL, consisting of a beam forming the seismic mass and flexible strips (tethers) acting as springs. The beam's displacement is monitored by a capacitive sensor consisting of interdigitated combs (framed part of Figures 6.15(a) and (b) and 6.16), the movable comb being integral to the seismic mass. The dimensions of the capacitive sensor are shown in Figure 6.15(b).

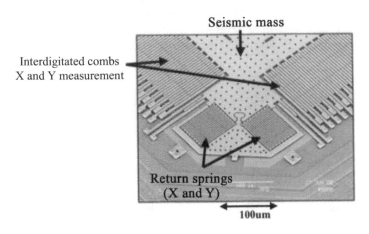

Figure 6.16. *Capacitive accelerometer with interdigitated combs with return springs (SEM) (Francais n.d.)*

6.4.4.2. *The gyroscope*

A gyroscope detects angular velocity, that is, the speed of rotation around an axis: roll, pitch and yaw, with angular velocities from 250°/s up to 2,000°/s.

Commercial gyroscopes measuring angular velocity are of the tuning fork type. They consist of two resonant "seismic mass-spring" structures set in vibration along a horizontal axis (drive axis) (Figure 6.17(a)).

This structure, moving along the x-axis, goes around the vertical z-axis (counter-clockwise), each mass being subjected to the Coriolis force generated by the rotation, which is exerted perpendicular to the direction of the movement of each mass and produces a vibration in the direction of the z-axis.

Figure 6.17(b) shows a tuning fork gyroscope and the lumped-element model for each element.

Vibratory displacements in mode 1 (x-direction) are electrostatically driven, and in mode 2 (y-direction) they are picked up by two interdigitated comb structures (STM 2011).

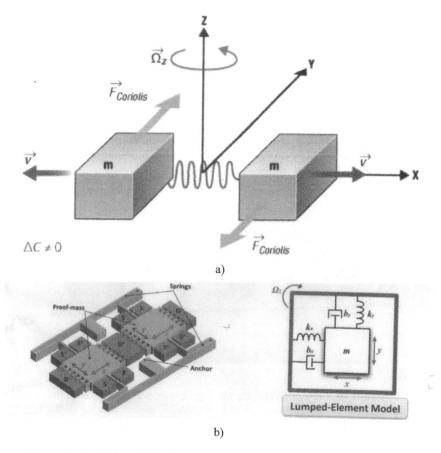

Figure 6.17. *(a) Principle of operation of a vibrating structure gyroscope; (b) tuning fork gyroscope (TFG) and vibration modes (Serrano 2013). For a color version of this figure, see www.iste.co.uk/vignes/silicon2.zip*

Figures 6.18 and 6.19 show two types of gyroscopes.

Figure 6.18. *Prototype of the Draper Lab comb drive tuning fork. The interdigitated comb actuator activates a tuning fork. One can see the two seismic masses, the pendulums (springs) and the interdigital comb structures (electrostatic actuator and capacitive sensor) (Burg et al. 2004)*

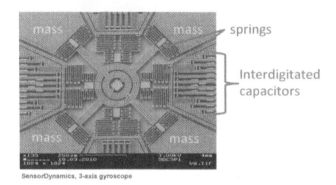

Figure 6.19. *Sensor Dynamics three-axis gyroscope. The structure measures 800 µm (Molinaro and Français 2014). For a color version of this figure, see www.iste.co.uk/vignes/silicon2.zip*

6.4.5. *Micromotors*

In the early 1990s, researchers at UC Berkeley (Long-Shen et al. 1989) and MIT (Mehregany and Tai 1991) independently developed electrostatically actuated rotating micromotors made entirely of monocrystalline silicon, featuring a rotor and bearing (Figure 6.20). Figure 6.20(a) shows such a motor. The central rotor carries an electrical load. The stator poles carry an opposite load. Each pair of diametrically opposed poles is successively and alternately polarized (on/off), causing the rotor to rotate. Figure 6.20(b) shows a cross-section of the motor showing bearing, rotor, bushing and stator elements. The gap between the rotor and the stator is 2.5 µm. Micromotors with diameters of around 100 µm running at speeds of 15,000 rpm were produced.

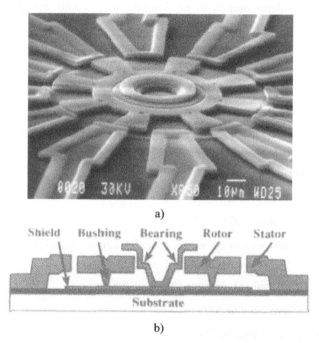

a)

b)

Figure 6.20. (a) SEM of a 100 µm diameter rotary electrostatic micromotor, micromachined by surface etching; (b) cross-section (Judy 2001)

Of the three families of micromotors (electrostatic, electrical induction, piezoelectric), electrostatic micromotors are the only ones that can be manufactured in silicon using conventional microelectronics and MEMS techniques, which explains their development.

The difficulty of integrating magnetic or piezoelectric materials into MEMS technology is the reason why induction and piezoelectric motors have not been developed (Liu et al. 2010).

However, these electrostatic micromotors require high electrical voltages to achieve high speeds and torques, which are incompatible with the voltages of mid-range electronics (<5 V). What is more, these motors, with their rubbing surfaces between bearing and rotor, are subject to wear that is incompatible with long-term use. Although there are no commercial products yet, the electrostatic micromotor is yet another example of the potential of MEMS.

The Silmach micromotor with an interdigitated comb actuator driving the rotation of a gearwheel was presented (Figure 6.11).

6.5. MEMS manufacturing technologies

Although using microelectronics technologies, the manufacture of these MEMS has required the development of specific technologies and lengthy studies and trials. The time taken to develop these MEMS, from initial research to commercialization, has in most cases been in excess of a decade (Lamers and Pruitt 2010), for example:

– pressure transducers from 1970 to 1980;

– inkjet printer heads from 1975 to 1985;

– accelerometers from 1975 to 1990;

– digital micromirrors and optical switches on certain models of video projectors from 1975 to 1997;

– microfluidic control valves from 1978 to 1998;

– gyroscopes from 1990 to 2002;

– microphones from 1983 to 2003.

A MEMS is a fixed part on a single-crystal silicon substrate and a moving part in polycrystalline silicon, a tiny mechanism (it could be a resonator, a micromotor, etc.).

Most of the work done on the substrate is based on the technologies used in microelectronics today: resin bonding, photolithography dry or wet etching. The main specificities of MEMS production technologies, in comparison with microelectronics, relate to the production of moving parts.

Two techniques have been developed.

6.5.1. Bulk micromachining

Two basic technologies are used as follows (see section 1.4.3.4 and Figure 1.21).

6.5.1.1. *Wet chemical etching*

Chemical etching consists of selectively etching the silicon substrate with a chemical agent (a base such as KOH) along a given crystallographic plane (depending on the chemical agent) to create holes, V-grooves or sensor membranes (Figure 6.21).

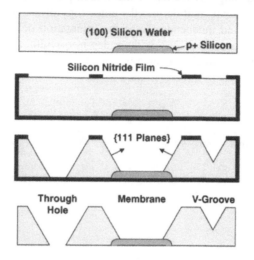

Figure 6.21. *Chemical etching of single-crystal silicon along crystallographic planes (Judy 2001)*

6.5.1.2. *Deep reactive ion etching*

DRIE (deep reactive ion etching) allows single-crystal substrates to be etched to thicknesses of several tens of micrometers, and even allows micromechanical parts with very high precision to be manufactured. It is particularly well suited to silicon, for which the maturity of this technology means that it is now possible to obtain highly vertical etchings, opening up new prospects for a material with singular mechanical properties (UFC 2011).

The substrate to be machined is placed in a vacuum chamber, into which an ionized gas (or plasma) is introduced. Under the effect of an electric field, the ions are set in motion, ensuring machining by mechanical effect, linked to the impact of the ions on the substrate. Reactive gases are also introduced; the reaction of these gases with the substrate material ensures an additional chemical effect. The type of material determines the nature of the gas used; in the case of silicon, fluorinated gases such as sulfur hexafluoride (SF_6) are preferred.

The process consists of successive etching and passivation: $SF_6 + O_2$ plasma etching and flank passivation by depositing a polymer layer on the etched flanks (Figure 6.22).

Vertical etching is accompanied, during the same cycle, by a fluorinated gas deposition phase, causing passivation of the silicon part's flanks as it is processed. This action is fundamental, as it defines the anisotropy of the etching along the axis of ion bombardment and independently of the orientation of the crystalline planes making up the material. Trenches with a depth-to-width ratio of 50:1 can be obtained, and even full-thickness etching of the silicon wafer is possible for watchmaking parts.

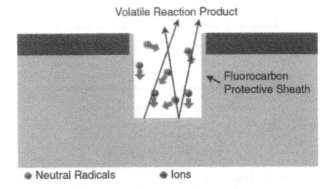

Figure 6.22. *DRIE etching: diagram showing the anisotropy of the etching and the protective layer on the sides of the trench (Tadigadapa and Lärmer 2011). For a color version of this figure, see www.iste.co.uk/vignes/silicon2.zip*

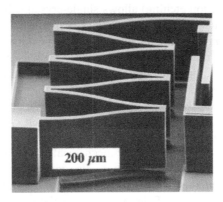

Figure 6.23. *Microflexure created by vertical etching through a wafer with DRIE (Judy 2001)*

DRIE can be used to produce highly complex profiles, enabling the manufacture of parts in a wide variety of shapes (Figure 6.23).

6.5.2. Surface micromachining

Surface micromachining is based on the presence of two types of layers: a sacrificial layer and a structural layer.

The sacrificial layer is an oxide (PSG: phosphosilicate glass).

The structural layer is made of polysilicon.

At the end of the process, the sacrificial layer is removed by selective, isotropic etching (using a buffered HF bath).

The steps involved in manufacturing an anchor stud by surface micromachining of polycrystalline silicon are shown in Figure 6.24 (Judy 2001; Howe and Muller 1986).

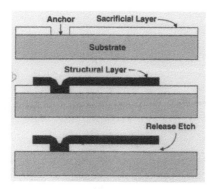

Figure 6.24. *Manufacture of a beam by surface micromachining and the sacrificial layer technique of silicon (Judy 2001)*

6.5.3. Manufacturing processing sequence of a lateral resonant structure

The main surface and volume micromachining steps in the manufacturing process of a structure made up of a fixed structure (anchor + comb finger) and a free suspended plate are presented in Figure 6.25.

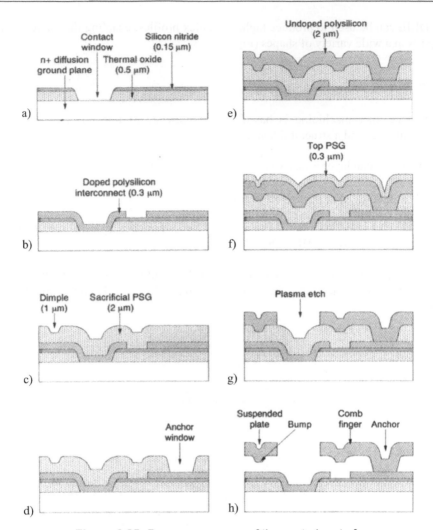

Figure 6.25. *Process sequence of the central part of a lateral resonant structure (Tang 1990, p. 67, figure 3.1)*

The manufacturing process consists of the following steps:

– Step a: this includes the opening of a "contact window" in the thermal oxide layer using a combination of reacting-ion-etching (RIE) in an SF_6 plasma and wet etching in a 5.1 buffered HF bath.

– Step b: this includes deposition of the first doped polysilicon layer by LPCVD and opening a window in this layer in a CCl_4-O_2 plasma.

– Step c: this includes deposition of (a) LPCVD sacrificial phosphosilicate glass (PSG), and (b) dimples are formed in a CHF_3-CF4 plasma.

– Step d: this includes the opening of an anchor window in the sacrificial layer using RIE and wet etching.

– Step e: this includes deposition of an undoped polysilicon structural layer by LPCVD.

– Step f: this includes deposition of a PSG layer.

– Step g: this includes stripping of the top PSG layer in buffered HF and the opening of a window in the undoped polysilicon layer by RIE in a CCl_4 plasma.

– Step h: this includes removing the sacrificial PSG layer in a 5.1 buffered HF to produce the moving and suspended parts of the device.

6.6. References

Adell, M. (2003). Surface mount microphones. White Paper, Knowles Acoustics, 1–8.

Burg, A., Meruani, A., Sandheinrich, B., Wickmann, M. (2004). MEMS gyroscopes and their applications. ME 381. Introduction to microelectromechanical system [Online]. Available at: https://media.gradebuddy.com › documents.

Chaehoi, A. (2005). Conception et modélisation de MEMS monolithique CMOS en technologie FSBM : application aux accéléromètres. PhD Thesis, Université de Montpellier, Montpellier.

Chen, P., Muller, R.S., Jolly, R.D., Halac, G.L., White, R.M., Andrews, A.P., Lim, T.C., Mohamedi, M.E. (1982). Integrated silicon microbeam PI-FET accelerometer. *IEEE Transaction on Electron Devices*, ED-29, 27–33.

Clarke, S.K. (1979). Pressure sensitivity in anisotropically etched thin diaphragm pressure sensors. *IEEE Trans. Electron Devices*, 26, 1886–1896.

Français, O. (n.d.). Qu'est-ce qu'un MEMS ? [Online]. Available at: http://www.esiee.fr/-francaio.

Frantz, J. (1988). Aufbau Funktionsweise und technische Realisierung eines piezoelektrischen Siliciumsensors für akustische Grössen. *VDI- Berichte*, 667, 299–302.

Hohm, D. and Sesssler, G.M. (1983). An integrated silicon-electret condenser microphone. In *11th ICA Paris Lyon Toulouse*, 29–32 [Online]. Available at: http://eduscol.education.fr/sti/si-ens-cachan.

Howe, R.T. and Muller, R.S. (1986). Resonant-microbridge vapor sensor. *IEEE Transaction on Electron Devices*, 33(4), 499–506.

Institut Pierre Vernier (2011). Technologie DRIE et silicium. *En direct*, 237(7).

Judy, J.W. (2001). Micromechanical systems (MEMS). *Smart Materials and Structures*, 10, 1115–1134.

Lamers, T.L. and Pruitt, B.L. (2010). The MEMS process design. In *MEMS Materials and Processes Handbook*, Ghodssi, R. and Lin, P. (eds). Springer.

Liu, D., Friend, J., Yeo, L. (2010). A brief review of actuation at the microscale using electrostatics, electromagnetics and piezoelectric ultrasonics. *Accoustic Science and Technology*, 31(2), 115–123.

Long-Shen, F., Tai Yu-Chong, T., Muller, R.S. (1989). IC-processed electrostatic micromotors. *Sensors Actuators*, 20, 41–47.

Man, D. (n.d.) Transducteurs et capteurs. EPFL [Online]. Available at: https://dokumen.tips.

Marcillat, G. (1985). Capteur accélerométrique à structure pendulaire plane. Patents, FR8403441, 0157663A1, US4663972A.

McWhan, D. (2012). *Sand and Silicon: Science that Changed the Word*. Oxford University Press, Oxford.

Mehregany, M. and Senturia, S.D. (1992). Micromotor fabrication. *IEEE Transactions on Electron Devices*, 39(9), 2060–2069.

Mehregany, M. and Tai, Y.-C. (1991). Surface micromachined mechanisms and micromotors. *J. Micromechanics and Microengineering*, 1, 73–85.

Molinaro, H.H. and Français, O. (2014). Les technologies MEMS [Online]. Available at: http://eduscol.education.fr/sti/si-ens-cachan.

Molinaro, H.H. and Français, O. (2015). Réalisation de capteurs résonants [Online]. Available at: http://eduscol.education.fr/sti/si-ens-cachan/2.

Pfann, W.G. (1962). Isotropically piezoresistive semiconductor. *Journal of Applied Physics*, 33, 1618–1619.

Pfann, W.G. and Thurston, R.N. (1961). The MEMS process design. Semiconducting stress transducers utilizing the transverse and shear piezoesistance effects. *Journal of Applied Physics*, 32, 2008–2019.

Polla, D.L. and Francis, L.F. (1998). Processing characterization of piezoelectric materials and integration into micromechanical systems. *Ann. Rev. Mater. Sci.*, 28, 563–597.

Prime Faraday Technology Watch (2002). An introduction to MEMS [Online]. Available at: www.primefaradaytechnologywatch.org.uk.

Royer, M., Holmen, J.O., Wurm, M.A., Aadland, O.S., Glenn, M. (1983). ZnO on Si integrated acoustic sensor. *Sensors and Actuators*, 4, 357–362.

Roylance, L.M. and Angell, J.B. (1979). A batch-fabricated silicon accelerometer. *IEEE Transaction on Electron Devices*, ED-26, 1911.

Scheeper, P., Gullov, J.O., Kofoed, L.M. (1996). A piezoelectric triaxial accelerometer. *Journal of Micromechanics and Microengineering*, 6(1), 131.

Schellin, R. and Hess, G. (1992). A silicon subminiature microphone based on piezoresistive polysilicon strain gauges. *Sensors and Actuators*, A32, 555–559.

Serrano, D.E. (2013). Design and analysis of MEMS gyroscopes. *IEEE Sensors* [Online]. Available at: www.ieee-sensors2013.org.

Silmach (2009). Micromoteurs "horlogers". Patents, US7636277, US8058772, US0070069604.

Smith, C.S. (1954). Piezoresistivity effect in Ge and Si. *Physical Review*, 94, 42–49.

STM (2011). STMicroelectronics. Everything about STMicroelectronics' 3-axis digital MEMS gyroscopes. TA0343 [Online]. Available at: www.st.com/resource/en/applicationnote/dm00103199.pdf.

STM (2017). Tutorial for MEMS microphones. STMicroelectronics.

Tadigadapa, S. and Lärmer, F. (2011). Dry etching for micromachining applications. In *MEMS Materials and Processes Handbook*, Ghodessi, R. and Lin, P. (ed.). Springer, Berlin.

Tang, W.C. (1990). Electrostatic comb drive for resonant sensor and actuators applications. PhD Thesis, University of California, Berkeley.

Tang, W.C., Nguyen, T.C.H., Howe, R.T. (1989). Laterally driven polysilicon resonant microstructures. *Sensors and Actuators*, 20, 25–32.

Tufte, O.N. and Stelzer, E.L. (1963). Piezoresistive properties of silivcon. *Journal of Applied Physics*, 34, 313–318.

Tufte, O.N., Chapman, P.W., Long, D. (1962). Silicon diffused-element piezoresistive diaphragms. *Journal of Applied Physics*, 33, 3322.

Turner, G. (2004). History of gyroscopes [Online]. Available at: http://www.gyroscopes.org/history.asp.

UFC (2011). Technologie DRIE et silicium, une heureuse association. *En direct*, 237.

Velay, B. (2010). Modélisation d'un accéléromètre MEMS. *Le Bup*, 920, 3–24.

Vigna, B. (2013). La révolution des MEMS. *Paris Innovation Review*.

Wang, W.-C. (2006). Introduction to microsensors and microactuators [Online]. Available at: http://depts.washington.edu/microtech/optics/sensors/index.html.

Index of Names

B, C, D

E, F, H

J, K

L, M

N, O, P

Index of Terms

A, B

accelerometer, 115, 128–130, 134
amplifier
  differential, 21–23
  operational, 20, 21, 23, 24
analog circuits, 20
bistable flip-flop, 6, 9, 18, 37, 40, 41

C

cadmium telluride
  CdTe, 63
capacitive detection, 120
capacitor/capacity
  depletion layer, 79, 80
  MOS, 44, 98, 99, 102, 104, 106
cell(s)
  DRAM, 35, 36, 43–47
  liquid crystal, 65
  ROM, 47
  solar
    amorphous silicon, 60
    CdTe, 74
    Cu_2S/CdS, 74
    thin-film, 71
  SRAM, 37, 38
  thick monocrystalline, 69

circuit(s)
  analog, 1, 20
  DTL, 15, 16, 18
  logic, 1, 2, 17, 19
  RC, 4
  RTL, 15–18
  TTL, 15, 18, 19
copper oxide Cu_2O, 58, 59

D, E

diffusion length, 81, 85–87
effect
  capacitive, 115, 122
  direct field, 57–60
  electrostatic, 122
  gettering, 79, 85, 87–89
  photoelectric, 52, 71, 98
  photoemissive, 68
  piezoelectric, 128
  piezoresistive, 111, 121
  tunneling, 51, 52
efficiency
  quantum, 68, 78, 99, 100, 102, 107
  solar conversion, 67, 70, 74, 84–86, 93

Summary of Volume 1

Preface

Introduction: The Digital Revolution

Chapter 1. Silicon and Germanium: From Ore to Element

Chapter 2. The Point-Contact Diode

2.1. Features and functions
    2.1.1. Characteristic curve, rectifier effect
    2.1.2. Rectifier contact and ohmic contact
    2.1.3. Point contact diode functions
    2.1.4. Physical basis for the operation of a point contact diode
2.2. History
    2.2.1. Discovery of the "rectifier effect"
    2.2.2. Discovery of AM radio wave "detection" by the
    point-contact diode
    2.2.3. Discovery of the silicon point-contact diode
    2.2.4. The germanium point-contact diode
    2.2.5. Reception of radar waves
2.3. Research during the Second World War
    2.3.1. Research on silicon
    2.3.2. Research on germanium
2.4. The industrial development of germanium diodes after the
Second World War
2.5. Appendix: currents in a metal–semiconductor diode
2.6. References

Chapter 3. The Point-Contact Transistor

3.1. The field effect
    3.1.1. "Direct" field effect, "inverse" field effect
    3.1.2. Bell Labs studies
3.2. The germanium-based point-contact transistor
    3.2.1. The discovery of the germanium N point-contact transistor
    3.2.2. Operation of the germanium-based N-tip transistor
    3.2.3. The point-contact transistor by Herbert Mataré and
    Heinrich Welker
3.3. The industrial development of the germanium
N point-contact transistor
3.4. References

Chapter 4. The PN Diode

4.1. PN diode operation and functions
    4.1.1. The discovery of the rectifier effect of the silicon-based
    PN diode
    4.1.2. PN diode operation
    4.1.3. PN diode functions

Chapter 6. The MOSFET Transistor

Other titles from

in

Materials Science

HICHER Pierre-Yves
Multiscale Geomechanics: From Soil to Engineering Projects

IONESCU Ioan R., BOUVIER Salima, CAZACU Oana, FRANCIOSI Patrick
Plasticity of Crystalline Materials: From Dislocations to Continuum

PIJAUDIER-CABOT Gilles, DUFOUR Frédéric
Damage Mechanics of Cementitious Materials and Structures

PRIESTER Louisette
Grain Boundaries and Crystalline Plasticity

VIGNES Alain
Extractive Metallurgy 1: Basic Thermodynamics and Kinetics
Extractive Metallurgy 2: Metallurgical Reaction Processes
Extractive Metallurgy 3: Processing Operations and Routes

2010

BATHIAS Claude, PINEAU André
Fatigue of Materials and Structures: Fundamentals

CHATEIGNER Daniel
Combined Analysis

CHEVALIER Yvon, VINH TUONG Jean
Mechanical Characterization of Materials and Wave Dispersion
Mechanical Characterization of Materials and Wave Dispersion:
Instrumentation and Experiment Interpretation

DELHAES Pierre
Carbon-based Solids and Materials

2009

ALVAREZ-ARMAS Iris, DEGALLAIX-MOREUIL Suzanne
Duplex Stainless Steels

DAVIM J. Paulo
Machining Composite Materials

GALERIE Alain
Vapor Surface Treatments

2008

BATTAGLIA Jean-Luc
Heat Transfer in Materials Forming Processes

BLONDEAU Régis
Metallurgy and Mechanics of Welding

FRANÇOIS Dominique
Structural Components

REYNE Maurice
Plastic Forming Processes

TAKADOUM Jamal
Materials and Surface Engineering in Tribology

2007

BOCH Philippe †, NIÈPCE Jean-Claude
Ceramic Materials

CRISTESCU Constantin
Materials with Rheological Properties

2006

LASSEN Tom, RECHO Naman
Fatigue Life Analyses of Welded Structures

Printed and bound by CPI Group (UK) Ltd, Croydon, CR0 4YY

27/10/2024

14580731-0001